Summary and Practice

with answers

Paul Hogan
St Wilfrid's C of E High School, Blackburn

Barbara Job
Christleton County High School, Chester

Text © Paul Hogan, Barbara Job 2002
Original illustrations © Nelson Thornes Ltd 2002

The right of Paul Hogan and Barbara Job to be identified as author of this work has been asserted by them in accordance with the Copyright, Designs and Patents Act 1988.

All rights reserved. No part of this publication may be reproduced or transmitted in any form or by any means, electronic or mechanical, including photocopy, recording or any information storage and retrieval system, without permission in writing from the publisher or under licence from the Copyright Licensing Agency Limited of 90 Tottenham Court Road, London W1T 4LP.

Any person who commits any unauthorised act in relation to this publication may be liable to criminal prosecution and civil claims for damages.

Published in 2002 by:
Nelson Thornes Ltd
Delta Place
27 Bath Road
CHELTENHAM
GL53 7TH
United Kingdom

02 03 04 05 / 10 9 8 7 6 5 4 3 2 1

A catalogue record for this book is available from the British Library.

ISBN 0 7487 6739 8

Page make-up by Tech Set Ltd

Printed and bound in Spain by Graficas Estella

Acknowledgements
The publishers thank the following for permission to reproduce copyright material:
Leslie Garland Picture Library 80; Collections/Brian Shuel 68; Martyn Chillmaid 112, 145, 147.
Thanks to Graham Newman, Chief Marker for Key Stage 3 Mathematics, for the Practice Examination Questions and Test Tips and for his invaluable contributions at review.

Contents

| How to use this book | 4 |

Number

1 Whole numbers	6
2 Fractions, decimals and percentages	16
3 Estimating	30

Algebra

4 Patterns and sequences	40
5 Formulas, expressions and equations	48
6 Functions and graphs	60

Shape, space and measures

7 2D and 3D shapes	72
8 Position and movement	86
9 Units of measurement	100
10 Perimeter, area and volume	110

Handling data

11 Organising data	120
12 Averages and spread	134
13 Probability	148

Test Yourself Answers	158
Practice Question Answers	168
Test Tips	176

How to use this book

This book has been carefully designed to help you to prepare for the National Key Stage 3 Tests that all students do in Year 9.

You will take the tests at one of four 'tiers'; Tier 3–5, Tier 4–6, Tier 5–7 or Tier 6–8. The numbers in these tiers are the National Curriculum levels. They are an indication of the difficulty of the work.

This book will help you to revise and give you practice for all of the tiers. You need to know which tier you are aiming for before you start using the book.

The book covers the National Curriculum in 13 short clear topics. On the next page, you can see how the book has been designed to help you study and practise your Maths. You should read this carefully before you start to use the book.

There are lots of Practice Questions. These are like the questions that you will get on the maths tests. One of the best ways to revise maths is to do lots of practice questions and check your answers.

When answering the questions you can write your answers in the spaces or write them on a sheet of paper. Your teacher will tell you which method to use.

The answers to all the questions are given at the back of the book.

Don't just copy the answers out. That won't help you learn! Try the questions and then check your answers. You can learn a lot by checking your own answers and looking to see where you may have lost marks, so that you can do better next time.

We hope you will find this book very helpful in preparing yourself for the tests.

Paul Hogan
Barbara Job

There are 13 topics in this book. Each one starts with work that **everyone needs to know**.

If you are doing Tier 6–8 or Tier 5–7 you may feel that you don't need to do any of the early part of each chapter. It is still a good idea to read it and do the Test Yourself and Practice Questions to make sure that you understand everything!

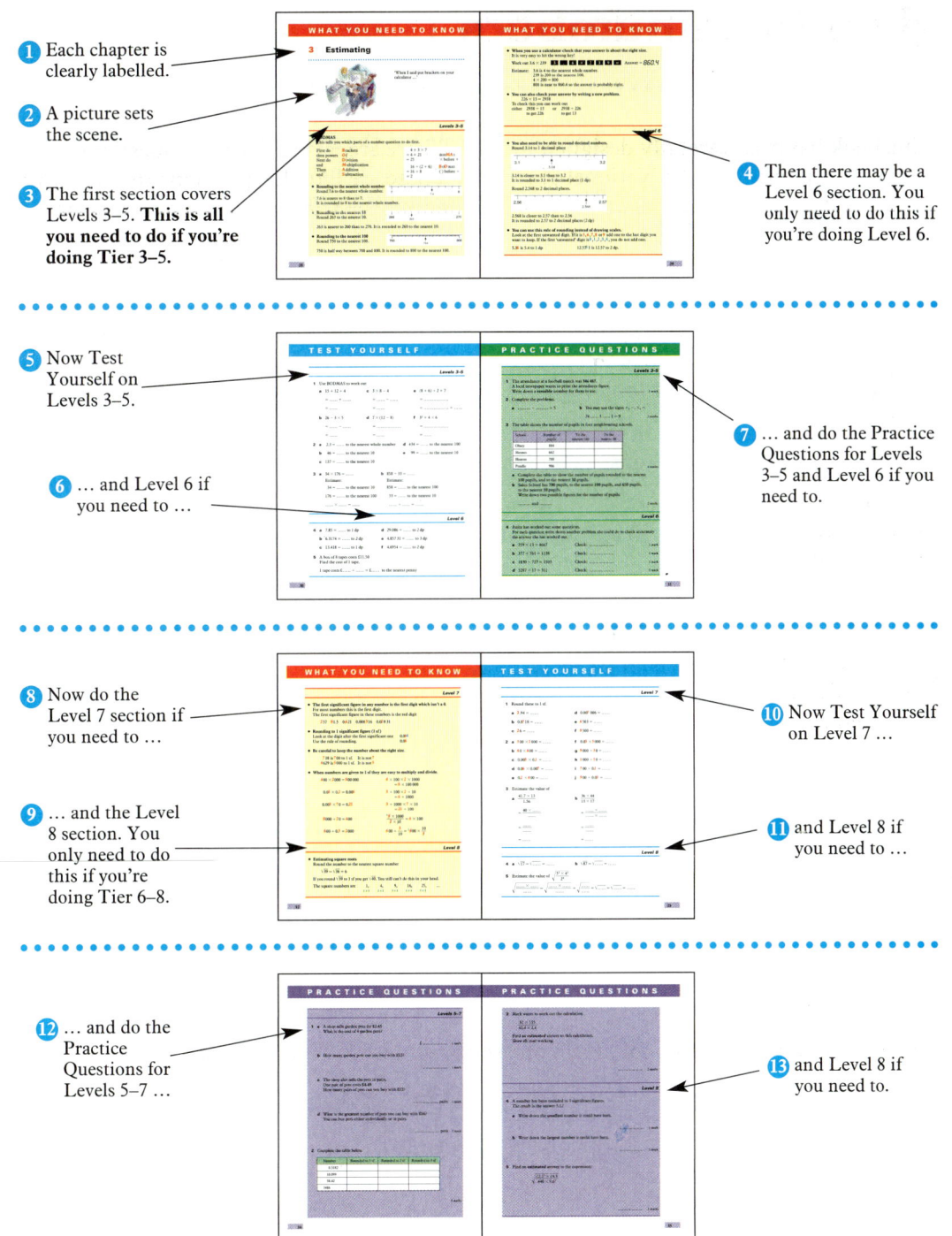

1. Each chapter is clearly labelled.

2. A picture sets the scene.

3. The first section covers Levels 3–5. **This is all you need to do if you're doing Tier 3–5.**

4. Then there may be a Level 6 section. You only need to do this if you're doing Level 6.

5. Now Test Yourself on Levels 3–5.

6. … and Level 6 if you need to …

7. … and do the Practice Questions for Levels 3–5 and Level 6 if you need to.

8. Now do the Level 7 section if you need to …

9. … and the Level 8 section. You only need to do this if you're doing Tier 6–8.

10. Now Test Yourself on Level 7 …

11. and Level 8 if you need to …

12. … and do the Practice Questions for Levels 5–7 …

13. and Level 8 if you need to.

5

WHAT YOU NEED TO KNOW

1 Whole numbers

The Celsius scale has the freezing point of water at 0 °C.
Temperatures below 0 °C have minus signs in front.
The temperature in this freezer is −10 °C.

Levels 3–5

- You need to know your tables up to the 10 times table.

- **The position of a digit in a number is important.**
  The 7 in 571 is worth 7 tens = 70
  The 7 in 4760 is worth 7 hundreds = 700

  Th H T U
     5 7 1
  4 7 6 0

- **When you multiply by 10 all the digits move one column to the left.**
  The number gets bigger.

  H T U
    4 6
  4 6 0
  46 × 10 = 460

- **When you divide by 10 all the digits move one column to the right.**
  The number gets smaller.

  H T U
  5 8 0
    5 8
  580 ÷ 10 = 58

- **When you multiply or divide by 100 the digits move two columns instead of one.**

- **When you multiply or divide by 1000 the digits move three columns.**

  Th H T U
  2 3 0 0
      2 3
  2300 ÷ 100 = 23

- **Addition and subtraction without a calculator**
  Put the digits in the correct columns.
  In this question 8 + 5 = **13**
  Put the **3** in the units column and carry the **1**.

  H T U
    2 4 8
  +   2 5
    2 7 3
       1

- **You may have to borrow in subtractions.**
  You cannot take 9 from 6. You borrow a ten to give
  16 − 9 = 7

  H T U
  2 ²3̸ ¹6
  −   1 9
  2 1 7

6

WHAT YOU NEED TO KNOW

- **Multiplication without a calculator**
  Multiply by the units digit first.
  Put 0 in the units column before multiplying
  by the tens digit.
  Add the two parts.

  $$\begin{array}{r} 1\ 2\ 3 \\ \times\ \ \ 3\ 2 \\ \hline 2\ 4\ 6 \\ 3\ 6\ 9\ 0 \\ \hline 3\ 9\ 3\ 6 \\ 1 \end{array}$$

- **Division without a calculator**
  In this question start by working out the
  13 times table.
  $31 \div 13 = 2$ with 5 to carry because
  $2 \times 13 = 26$ and $31 - 26 = 5$
  Then $52 \div 13 = 4$ because $4 \times 13 = 52$

  $13\ )\overline{3\ 1^5 2}$ with quotient $2\ 4$

  $1 \times 13 = 13$
  $2 \times 13 = 26$
  $3 \times 13 = 39$
  $4 \times 13 = 52$

- **Negative numbers have a minus sign in front of them.**
  This thermometer shows a temperature of $-5\,°C$
  $-10\,°C$ is lower than $-5\,°C$.
  $5\,°C$ is higher than $-5\,°C$.

- **Use a thermometer scale to help you add and subtract.**
  To work out $-5 + 8$ start at -5 and move up 8.
  You get to 3   so $-5 + 8 = 3$
  To work out $-5 - 2$ start at -5 and move down 2.
  You get to -7 so $-5 - 2 = -7$

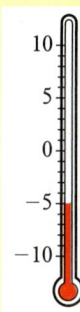

Level 6

- **Adding and subtracting with negative numbers.**
  $6 - -3 = 6 + 3 = 9$ $--$ becomes $+$
  $7 + -2 = 7 - 2 = 5$ $+-$ becomes $-$
  $-4 - -5 = -4 + 5 = 1$
  $-2 + -6 = -2 - 6 = -8$

- **Multiplying and dividing with negative numbers.**
  $+ \times + = +$ $- \times - = +$ Two signs the same give $+$
  $+ \times - = -$ $- \times + = -$ Two different signs give $-$
  The same rules work for division.
  $+4 \times +3 = +12$ $-4 \times -6 = +24$ $+12 \div +6 = +2$ $-24 \div -6 = +4$
  $+5 \times -2 = -10$ $-10 \times +3 = -30$ $+20 \div -5 = -4$ $-16 \div +2 = -8$

- 5^3 means $5 \times 5 \times 5$. The power 3 tells you how many 5s to multiply together.

- **A power of 2 is called square.**
  10^2 means 10×10
  You say this as 'ten squared'.

- **A power of 3 is called cube.**
  10^3 means $10 \times 10 \times 10$
  You say this as 'ten cubed'.

7

TEST YOURSELF

Levels 3–5

1 The value of

Th	H	T	U
5	1	3	8

   a the 3 is ......

   b the 1 is ......

2 **a** $4 \times 7 =$ ......   **b** $6 \times 9 =$ ......   **c** $8 \times$ ...... $= 24$

3 **a** $26 \times 10 =$ ......   **b** $41 \times 100 =$ ......   **c** $53 \times$ ...... $= 530$

4 **a** $360 \div 10 =$ ......   **b** $8000 \div 100 =$ ......   **c** $4860 \div$ ...... $= 486$

5 **a** 260 **b** 187 **c** 174 **d** 509
        $+\,123$ $+\ \,35$ $-\ \,51$ $-\ \,46$

6 **a** 237 **b** $14\overline{)322}$ $1 \times 14 =$
        $\times\ \,25$ $2 \times 14 =$
                                           $3 \times 14 =$

7 Put these numbers in order $12, -4, -6, 5, 0$

   Start with the smallest   ...... , ... , ... , ... , ...

8 The temperature is $-2\,°C$. It rises $5\,°C$. The new temperature is ...... $°C$.

Level 6

9 **a** $4 - {-6} =$ ......   **d** $20 \div {-4} =$ ......   **g** $-5 \times$ ...... $= 15$

   b $-8 - 3 =$ ......   **e** $18 - {-2} =$ ......   **h** $-36 \div$ ...... $= -6$

   c $-4 \times -7 =$ ......   **f** $-6 + {-12} =$ ......   **i** ...... $\div -10 = -5$

PRACTICE QUESTIONS

Levels 3–5

1 Here are some number cards.

| 4 | 2 | 3 | 5 |

    a Write down the **smallest** number that can be made with these cards.

                                                                                     .... 1 mark

    b Write down the **largest** number that can be made with these cards.

                                                                                     .... 1 mark

    c Which extra card is needed to make a number **ten** times bigger than **4235**?

                                                                                     .... 1 mark

2 a A shop sells CDs for **£13** each.
Find the cost of **25 CDs**.
Show your working.

                              Cost of 25 CDs is £ .... 2 marks

    b The shop also sells packs of video tapes for **£18**.
What is the greatest number of packs of video tapes that could be bought with **£300**?
Show your working.

                                                               .... packs   2 marks

PRACTICE QUESTIONS

3 a Write down the **temperature** shown by the arrow.

.... °C   1 mark

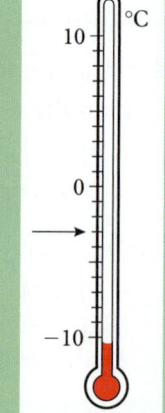

b Draw an arrow on the scale to show a temperature of **8 °C**.

1 mark

c A temperature of **4 °C** went down **6 °C**.
What is the temperature now?

.... °C   1 mark

Level 6

4 −5, −4, −3, −2, −1, 0, 1, 2, 3, 4, 5

Choose a number from the list above which will give

a the **lowest** possible answer

−4 − .... = ...   1 mark

b the **highest** possible answer

−4 − .... = ...   1 mark

In each case also work out the answer.

5 Substitute the values $a = 15, b = -2$ into the formula and work out the value of T:

$$T = \frac{4a}{b}$$

$T = $ ....   1 mark

10

WHAT YOU NEED TO KNOW

Level 7

- When you multiply by a number between 0 and 1 the answer will be smaller than the number you started with.

    $200 \times 0.5 = 100$ 100 is less than 200
    $300 \times \frac{2}{3} = 200$ 200 is less than 300

- When you divide by a number between 0 and 1 the answer will be larger than the number you started with.

    $200 \div 0.5 = 400$ 400 is more than 200
    $300 \div \frac{2}{3} = 450$ 450 is more than 300

- Use the bracket keys ( ) on your calculator to help you work out complicated problems.

    To work out $\dfrac{246 - 157}{4.3 \times 9.1}$ put brackets around the top and bottom. Key in

    (2 4 6 − 1 5 7) ÷ (4 . 3 × 9 . 1) =

    to get 2.27 to 2 dp.

- The opposite of 'square' is 'square root'.
    $\sqrt{25} = 5$ because $5 \times 5 = 25$
    -5×-5 is also 25 so $\sqrt{25}$ can be 5 or -5.

- The opposite of 'cube' is 'cube root'.
    You write the cube root as $\sqrt[3]{}$
    $\sqrt[3]{8} = 2$ because $2 \times 2 \times 2 = 8$

Level 8

- **Power rules**
    When you multiply, add the powers. $a^3 \times a^5 = a^{3+5} = a^8$
    When you divide, subtract the powers. $b^7 \div b^4 = b^{7-4} = b^3$
    With brackets, multiply the powers. $(c^4)^5 = c^{4 \times 5} = c^{20}$

- **Negative powers**
    When you get a negative power it always means divide,
    2^{-3} means divide by 2^3.

    So $2^{-3} = 1 \div 2^3 = \dfrac{1}{2^3} = \dfrac{1}{8}$ and $10^{-4} = \dfrac{1}{10^4} = \dfrac{1}{10\,000}$

WHAT YOU NEED TO KNOW

- **Any number can be written as a number between 1 and 10 multiplied by a power of 10.** This is called standard form. To write 67 000 in standard form, the number between 1 and 10 is 6.7. You need to multiply 6.7 by 10 four times to get 67 000.

    So 67 000 = 6.7 × 10⁴   6.7 0 0 0. (1 2 3 4)

    To write 0.000 024 6 in standard form, the number between 1 and 10 is 2.46 You need to divide by 10 five times to get to 0.000 024 6.

    To show you have to divide, put a minus sign in the power.   0.0 0 0 0 2.4 6 (5 4 3 2 1)

    So 0.000 024 6 = 2.46 × 10⁻⁵

- **You use these rules to change numbers in standard form back to ordinary numbers.**

    7.13 × 10⁶ = 7 130 000   7.1 3 0 0 0 0. (1 2 3 4 5 6)

    4.02 × 10⁻³ = 0.004 02   0.0 0 4.0 2 (3 2 1)

- Most calculators have an **EXP** or **EE** key.
    You can use this key to enter numbers in standard form.

    For 7.13 × 10⁶ key in **7** **.** **1** **3** **EXP** **6**

    The calculator display will look like 7.13×10^{06}

    Use the **+/−** key to enter negative powers.

    For 4.02 × 10⁻³ key in **4** **.** **0** **2** **EXP** **+/−** **3**

    The calculator display will look like 4.02×10^{-03}

- **Never write the calculator display as your answer.**

    6.5×10^{-04} must be written 6.5 × 10⁻⁴ with the 10 shown full size.

    To work out (3.9 × 10⁴) ÷ (1.3 × 10⁻⁶) key in

    3 **.** **9** **EXP** **4** **÷** **1** **.** **3** **EXP** **+/−** **6** **=**

    to get $3. \times 10^{10}$. You must write this as 3 × 10¹⁰.

TEST YOURSELF

Level 7

1 Fill in the space with × or ÷

    a 60 …… 0.2 = 300   **c** 260 …… 0.4 = 104

    b 50 …… 0.3 = 15   **d** 150 …… 0.1 = 1500

2 Write down the calculator keys you would press to work out

    a $\dfrac{4.86 \times 1.63}{4.37 + 1.94}$ ……………………………………………………

    b $\dfrac{4.2 + 3.6^2}{4.91 - 2.87}$ ……………………………………………………

3 **a** $25^2 =$ ………………   **c** $\sqrt[3]{27} =$ ………………

    b the square root of 81 is ……   **d** the cube root of …… is 4

Level 8

4 **a** $p^5 \times p^4 =$ …………………   **c** $(r^3)^7 =$ ………………

    b $m^{12} \div m^7 =$ ……… = ………   **d** $4^{-3} =$ ……… = ………

5 Write these numbers in standard form

    a 0.005 = ………………   **b** 12 million = ………………

6 Write these as ordinary numbers

    a $2.81 \times 10^5 =$ ………………   **b** $8.3 \times 10^{-7} =$ ………………

7 Write down the calculator keys you would press to work out

    $(4.2 \times 10^6) \div (2.1 \times 10^{-4})$ ……………………………………………………

……………………………………………………………………………………………

PRACTICE QUESTIONS

Levels 5–7

1 Write numbers into each of the empty boxes so that each problem will work out correctly.

a 821 − 647 = ☐   d 182 ÷ 7 = ☐

b ☐ × ☐ × 5 = 40   e ☐ − ☐ = 31

c 1400 ÷ ☐ = 14   f 37 × 9 = ☐

6 marks

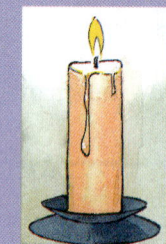

2 Sean makes large wax candles that are sold for **£27** each.

  a Sean sells **34** candles.
  How much does he get for the **34** candles?
  Show all your working clearly.

£ .... *2 marks*

  b Sean packs **12** candles in a box.
  He has **200** candles.
  How many full boxes can he pack with the **200** candles?
  Show all your working.

.... boxes *2 marks*

3 $K = \dfrac{14.2p}{q - r}$

  Use your calculator to find K when $p = 37.4$, $q = 80.4$, $r = 13.53$

$K = $ .... *1 mark*

PRACTICE QUESTIONS

4 The table shows information about the local taxes raised in four towns.

Town	Taxes	Population
A	£86 850 000	347 400
B	£119 945 000	521 500
C	£98 784 000	403 200
D	£170 977 500	670 500

a Which town had the lowest tax per person? .... 1 mark

b Which town had the highest tax per person? .... 1 mark

Level 8

5 Light travels at the speed of 9.46×10^{12} km/year.

The time it takes for light to reach Earth from the edge of the known universe is 15 000 000 000 years.

a Write the number 15 000 000 000 in standard form.

.... 1 mark

b Calculate the distance in kilometres from the edge of the known universe to Earth. Give your answer in standard form.

.... 2 marks

6 In the United States a centillion is 10^{303}.
In the United Kingdom a centillion is 10^{600}.
How many times bigger is a United Kingdom centillion than a United States centillion?

.... 2 marks

WHAT YOU NEED TO KNOW

2 Fractions, decimals and percentages

You would never see a price like this!

Levels 3–5

- **An amount of money which has both pounds and pence must have 2 decimal places.**
  £6.43 is £6 and 43 pence.

- **Your calculator sometimes only gives 1 decimal place.**
  If you work out £30 ÷ 4 on your calculator your display will look like this:
  You must write the answer as £7.50

  `7.5`

- **The position of a digit after a decimal point is important.**
  The value of the 3 is 3 **t**enths or $\frac{3}{10}$.
  The value of the 8 is 8 **h**undredths or $\frac{8}{100}$.

  ```
  U . t h
  4 . 3 8
  ```

- **Adding and subtracting decimals without a calculator.**
  Put the digits in the correct columns.
  In this question 7 + 6 = **13**
  Put the **3** in the 't' column and carry the **1**.

  ```
    T U . t h
    3 5 . 7 4
  + 1 2 . 6
    -----------
    4 8 . 3 4
          1
  ```

- **You may have to borrow in subtractions.**
  You cannot take 8 from 2 in the 't' column.
  You borrow one from the 'U' column to give 12 − 8 = 4

  ```
    T U . t h
    3 ⁴5̸ . ¹2 7
  − 1 3 . 8 5
    -----------
    2 1 . 4 2
  ```

- **Always make sure the decimal points go underneath each other.**

WHAT YOU NEED TO KNOW

- **When you multiply by 10 all the digits move one column to the left.**
  The number gets bigger.

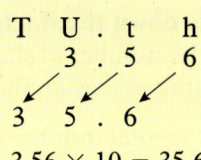

3.56 × 10 = 35.6

- **When you divide by 10 all the digits move one column to the right.**
  The number gets smaller.

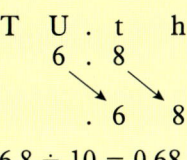

6.8 ÷ 10 = 0.68

- **When you multiply or divide by 100 the digits move two columns instead of one.**

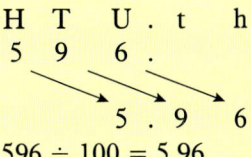

596 ÷ 100 = 5.96

- **When you multiply or divide by 1000 the digits move three columns.**

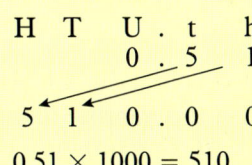

0.51 × 1000 = 510

- **Multiplying decimals without a calculator.**
  Start with the **4**.
  The $_2$ and the $_1$ are carries.
  Put 0 in the right hand column before multiplying by the **2**
  The $_1$ is a carry.
  Now add the two parts together.
  Make sure you don't add the carries.

$$\begin{array}{r} 1\,2\,.\,6\,3 \\ \times \quad\quad 2\,4 \\ \hline 5_2 0\,._1 5\,2 \\ 2_1 5\,2\,.\,6\,0 \\ \hline 3\,0\,3\,.\,1\,2 \end{array}$$

- **Dividing decimals without a calculator.**
  7 ÷ 5 = **1** with **2** to carry
  24 ÷ 5 = **4** with **4** to carry
  41 ÷ 5 = **8** with **1** to carry
  Add a **zero**
  10 ÷ 5 = **2**

$$\begin{array}{r} 1\,4\,.\,8\,2 \\ 5\overline{)7^2 4\,.^4 1^1 0} \end{array}$$

17

WHAT YOU NEED TO KNOW

- **To write down the fraction shaded**
  Count the number of shaded sections.
  This number goes on the top of the fraction.

  Count the total number of equal sections.
  This number goes on the bottom of the fraction.

  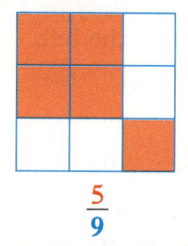

- **Different fractions can show the same amount.**

  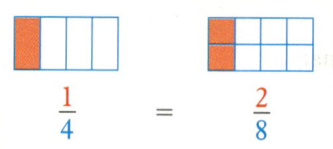 $\frac{1}{4}$ and $\frac{2}{8}$ are equivalent fractions

- **Cancelling fractions**

  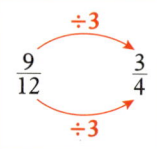   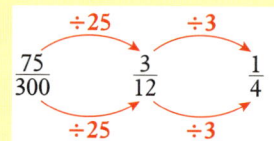

- **To find a fraction of an amount multiply by the top number and divide by the bottom number.**

  So $\frac{3}{5}$ of 400 g = 240 g

  $400 \times 3 = 1200$
  $1200 \div 5 = 240$

- **A percentage is a fraction where the bottom number is 100.**
  This circle has **100** equal sections.
  There are **15** shaded sections.

  The fraction shaded is $\frac{15}{100}$

  This is the same as **15%**

  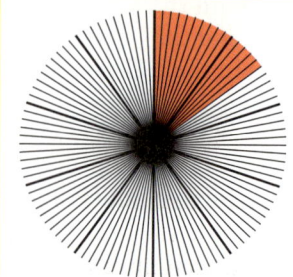

  - $\frac{1}{10} = 10\%$ $\frac{1}{5} = 20\%$ $\frac{1}{4} = 25\%$ $\frac{1}{2} = 50\%$
  - $\frac{7}{10} = 70\%$ $\frac{2}{5} = 40\%$ $\frac{3}{4} = 75\%$

- **To find a percentage of an amount multiply by the percentage and divide by 100.**
  So **65%** of £800 = £520

  $800 \times 65 = 52\,000$
  $52\,000 \div 100 = 520$

WHAT YOU NEED TO KNOW

Level 6

- **To put decimals in order of size**
  Look at the number before the decimal point first.

  26.415 is smaller than **40**.1 because **26** is smaller than **40**

  If the numbers before the decimal point are the same, look at the first number after the decimal point.

  6.**3**72 is smaller than 6.**4**3 because **3** is smaller than **4**

  Sometimes you need to look at the second number after the decimal point.

  7.2**0**8 is smaller than 7.2**6**1 because **0** is smaller than **6**

  Carry on like this. Look at one decimal place at a time.

- **To change a fraction to a decimal divide the top number by the bottom number.**

  $$\frac{7}{8} = 7 \div 8 = 0.875$$

- **To change a percentage to a fraction put it over 100.** Then you can get to a decimal by dividing if you need to.

  $$23\% = \frac{23}{100} = 0.23$$

- **To put a mixture of fractions, percentages and decimals in order, change them all to decimals first.**
  Remember to change them back at the end.

- **To change a decimal to a percentage multiply it by 100.
  To turn this into a fraction put it over 100.**

  $$0.31 = 0.31 \times 100\% = 31\%$$

  $$0.31 = 31\% = \frac{31}{100}$$

- You should be able to use the a^b/c key on your calculator to help you do fractions.

- **To write one number as a fraction of another number.**
  Make sure the units are the same.
  Put the first number on top of the second number.

  Use the a^b/c key to see if your fraction will cancel.

  To write 45 as a fraction of 180.

  Start with $\frac{45}{180}$. Key in **4** **5** a^b/c **1** **8** **0** **=**

  to get $\frac{1}{4}$ which is the simplest form of the fraction.

WHAT YOU NEED TO KNOW

- **To write one number as a percentage of another number**

    Write the numbers as a fraction.
    Change the fraction to a percentage by multiplying by 100.
    To write 26 as a percentage of 40

    Start with $\dfrac{26}{40} = \dfrac{26}{40} \times 100\% = 65\%$

- **Finding percentage change**

    The value of a car decreases from £10 000 to £8500 in one year.
    Find the percentage decrease.

    Percentage decrease = $\dfrac{\text{actual decrease}}{\text{original amount}} \times 100 = \dfrac{1500}{10\,000} \times 100 = 15\%$

- **Ratio can be used to compare two quantities.**

    There are 2 red counters   The ratio   red : blue
    and 4 blue counters   is   2 : 4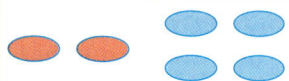

    You can simplify this because 2 goes into both numbers
    red : blue = 1 : 2

- **Ratio can also be used to show how a quantity is divided up.**

    To divide £70 in the ratio 2 : 3

    Find the total number of parts by
    adding the numbers in the ratio.   2 + 3 = 5 parts

    Find the value of one part.   £70 ÷ 5 = £14
    Work out the value of each share.   £14 × 2 = £28
                                                     £14 × 3 = £42

- **Scales on maps are often given as a ratio of the form 1 : n**

    Any ratio can be changed to this form.
    Divide both of the numbers in the ratio by the first number

    $$2 : 3 = \dfrac{2}{2} : \dfrac{3}{2} = 1 : 1.5$$

- **Sometimes ratios are given in the form n : 1**

    This time divide both numbers by the second number

    $$3 : 8 = \dfrac{3}{8} : \dfrac{8}{8} = 0.375 : 1$$

20

WHAT YOU NEED TO KNOW

- **Adding fractions**
  If the bottom numbers are the same add the top two numbers. $\frac{2}{7} + \frac{3}{7} = \frac{5}{7}$
  If the bottom numbers are the different make them the same. $\frac{2}{3} + \frac{1}{5}$

  3 and 5 both go into 15.

  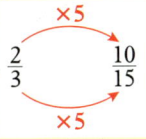 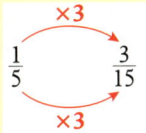

  Now you can add $\frac{2}{3} + \frac{1}{5} = \frac{10}{15} + \frac{3}{15} = \frac{13}{15}$

- **Subtracting fractions**
  This works just like adding fractions. $\frac{2}{3} - \frac{5}{12} = \frac{8}{12} - \frac{5}{12} = \frac{3}{12} = \frac{1}{4}$

- **Multiplying fractions**
  Multiply the two top numbers and $\frac{2}{3} \times \frac{3}{8} = \frac{6}{24} = \frac{1}{4}$
  multiply the two bottom numbers.
  Change any mixed numbers to improper fractions. $2\frac{1}{2} \times \frac{3}{8} = \frac{5}{2} \times \frac{3}{8} = \frac{15}{16}$

- **Dividing fractions**
  Turn the second fraction over and then multiply. $\frac{2}{3} \div 1\frac{3}{5} = \frac{2}{3} \div \frac{8}{5} = \frac{2}{3} \times \frac{5}{8} = \frac{10}{24} = \frac{5}{12}$

- **Powers of fractions**
  $(\frac{3}{5})^2$ means $\frac{3}{5} \times \frac{3}{5}$ so $(\frac{3}{5})^2 = \frac{3^2}{5^2} = \frac{9}{25}$

- **Ordering fractions**
  Think about moving along a number line to compare fractions.

  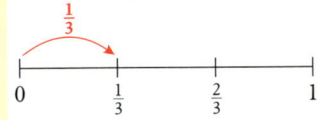 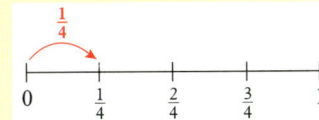 $\frac{1}{3}$ is bigger than $\frac{1}{4}$

  You can also think about moving back from the other end

  $\frac{1}{7}$ is smaller than $\frac{1}{6}$ so $\frac{1}{7}$ is a smaller step back from 1.

  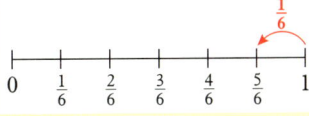 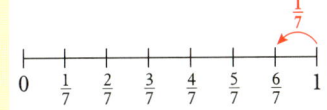 $\frac{6}{7}$ is bigger than $\frac{5}{6}$

  Another way to order fractions is to change them $\frac{3}{8}, \frac{5}{12}, \frac{1}{3}$
  so that they all have the same bottom number. $\frac{9}{24}, \frac{10}{24}, \frac{8}{24}$

  Now you can write the fractions in order. $\frac{1}{3}, \frac{3}{8}, \frac{5}{12}$

TEST YOURSELF

Levels 3–5

1 a   2.14
     + 1.79

  b   12.76
     − 8.58

  c   2.53
     ×   24

  d   5)8.95

2 a $\frac{5}{8}$ of £224 = £ ………

   …… ÷ …… = ………

   …… × …… = ………

  b 72% of 900 g = ……… g

   …… ÷ …… = ………

   …… × …… = ………

3 a Colour in 25%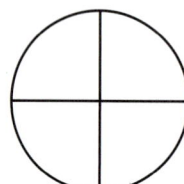

  b Colour in $\frac{3}{4}$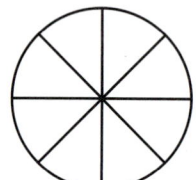

Level 6

4

Fraction	$\frac{1}{2}$	……	……	$\frac{3}{4}$	……	……
Decimal	……	0.25	……	……	0.6	……
Percentage	……	……	20%	……	……	81%

5 a 16 as a fraction of 40

   is $\dfrac{\text{……}}{\text{……}} = \dfrac{\text{……}}{\text{……}}$

  b 35p as a percentage of £2.50

   = 35p as a percentage of ……p = $\dfrac{\text{……}}{\text{……}} \times 100\%$

   = ……%

6 Share £180 in the ratio 2 : 3 : 4

  Total number of parts = ……

  Value of 1 part = …… ÷ …… = £ ……

  Value of each share

  2 × …… = £ ……   …… × …… = £ ……   …… × …… = £ ……

TEST YOURSELF

7 **a** $\frac{2}{5} + \frac{1}{5} = \frac{......}{......}$

   c $\frac{5}{7} - \frac{2}{7} = \frac{......}{......}$

   b $\frac{2}{3} + \frac{3}{4} = \frac{......}{......} + \frac{......}{......}$

   d $\frac{8}{9} - \frac{2}{3} = \frac{......}{......} - \frac{......}{......}$

   $= \frac{......}{......} = ...$

   $= \frac{......}{......}$

8 **a** $\frac{2}{7} \times \frac{3}{5} = \frac{......}{......}$

   d $\frac{3}{4} \div \frac{1}{8} = \frac{......}{......} \times \frac{......}{......}$

   b $\frac{3}{4} \times \frac{8}{9} = \frac{......}{......}$

   $= \frac{......}{......}$

   $= \frac{......}{......}$

   $= ......$

   c $2\frac{1}{3} \times 1\frac{1}{2} = \frac{......}{......} \times \frac{......}{......}$

   e $2\frac{1}{2} \div 1\frac{1}{3} = \frac{......}{......} \div \frac{......}{......}$

   $= \frac{......}{......}$

   $= \frac{......}{......} \times \frac{......}{......}$

   $= ......$

   $= \frac{......}{......}$

   $= ......$

9 Write these fractions in order starting with the smallest

   $\frac{1}{2}, \quad \frac{7}{15}, \quad \frac{3}{5}$

   $\frac{1}{2} = \frac{......}{......}$ $\frac{7}{15} = \frac{......}{......}$ $\frac{3}{5} = \frac{......}{......}$

   $\frac{......}{......}, \quad \frac{......}{......}, \quad \frac{......}{......}$

PRACTICE QUESTIONS

Levels 3–5

1 Kathy and Neil each have 16 counters.

    a Kathy uses **half** her counters for a game.
       How many counters does Kathy use?

.... 1 mark

    b Neil uses **4** counters for his game.
       What **fraction** of his counters did Neil use?

.... 1 mark

    c **How many** counters does Kathy have left?

.... 1 mark

    d **How many** counters does Neil have left?

.... 1 mark

2 Here are the ingredients for a Greek meal.
    The table shows the quantities needed for 6 people.

6 people	9 people
Cloves of garlic: 2	Cloves of garlic: ……
Chick peas: 4 ounces	Chick peas: …… ounces
Tablespoons olive oil: 4	Tablespoons olive oil: ……
Paste: 5 fluid ounces	Paste: …… fluid ounces

Complete the table to show the quantities needed for 9 people.

2 marks

PRACTICE QUESTIONS

3 a What **fraction** of the diagram is shaded?

.... 1 mark

b What **percentage** of the diagram is shaded?

.... 1 mark

c Shade $\frac{2}{3}$ of the second diagram.

1 mark

4 Kevin has bought a computer game from his uncle.
His uncle asks him for weekly payments which are either **5% of £30** or $\frac{2}{5}$ **of £4**.

a Work out 5% of £30.   £ .... 1 mark

b Work out $\frac{2}{5}$ of £4.   £ .... 1 mark

c Which is the least amount that Kevin will need to pay per week?   £ .... 1 mark

Level 6

5 To make a shade of orange paint, **2 parts red** is mixed with **3 parts yellow**.
12 litres of **yellow** paint are to be mixed with red paint to make some orange paint.

a How many litres of **red** paint are needed?

.... litres of red paint   2 marks

b What is the total quantity of **orange** paint made?

.... litres of orange paint   1 mark

WHAT YOU NEED TO KNOW

Level 7

- **To increase an amount by a given fraction or percentage**
  Work out the fraction or percentage. Add it on to the original amount.

  Increase £400 by 5%   5% of £400 = £20   New amount = £420

- **To decrease an amount by a given fraction or percentage**
  Work out the fraction or percentage. Take it away from the original amount.

  Reduce 375 g by $\frac{2}{15}$ $\frac{2}{15}$ of 375 g = 50 g   New amount = 325 g

- **You may be given harder fraction questions.**
  $3\frac{1}{3} - 1\frac{3}{4} = \frac{10}{3} - \frac{7}{4} = \frac{40}{12} - \frac{21}{12} = \frac{19}{12} = 1\frac{7}{12}$

Level 8

- **You may have to increase or decrease the amount more than once.**
  £100 is invested. Each year 6% is added as interest.
  Find the balance after 3 years.

Year 1	Interest = 6% of £100 = £6	New balance = £106
Year 2	Interest = 6% of £106 = £6.36	New balance = £112.36
Year 3	Interest = 6% of £112.36 = £6.74	New balance = £119.10

- **You may be given the final amount and how it has been changed. You will have to find the original amount.**
  A coat is reduced by **15**% in a sale. The sale price is £61.20
  Find the original price.
  The sale price is 100% − **15**% = 85% of the original price.
  So 1% of the original price = £61.20 ÷ 85 = £0.72
  100% of the original price = £0.72 × 100 = £72

- **Sometimes changes are given as fractions.**
  $\frac{5}{12}$ of a number is 45.
  Find the number.
  $\frac{5}{12}$ is 45 so $\frac{1}{12}$ is 45 ÷ **5** = 9.
  So $\frac{12}{12}$ = 9 × **12** = 108

- **Recurring decimals are exact fractions.**
  $0.\dot{1} = \frac{1}{9}, 0.\dot{2} = \frac{2}{9}, 0.\dot{3} = \frac{3}{9} = \frac{1}{3}$ and so on.

TEST YOURSELF

Level 7

1 Increase **a** 650 g by 20%   **b** £460 by $\frac{2}{5}$

  a 20% of …… = ……   **b** …… of …… = ……

   New amount = …… g   New amount = £ ……

2 Decrease **a** 370 cm by 15%   **b** 2400 kg by $\frac{3}{8}$

  a …… of …… = …… cm   **b** …… of …… = …… kg

   New amount = …… cm   New amount = …… kg

3 VAT at 17.5% is added to a bill of £60. Find the total.

   17.5% of …… = £ ……   Total bill = £ ……

Level 8

4 Asha buys a car for £8600. The value of the car decreases by 15% every year. How much is the car worth after 3 years?

   Year 1 Loss = 15% of …… = £ ……   New value = £ ……

   Year 2 Loss = …… of …… = £ ……   New value = £ ……

   Year 3 Loss = …… of …… = £ ……   New value = £ ……

5 The population of Plumley village has increased by 7% since 1990. It is now 2568. What was the population in 1990?

   New population = 100% + …… = …… % of the 1990 population

   So 1% of 1990 population = …… ÷ …… = ……

   100% of 1990 population = …… × 100 = ……

6 Every year a firm increases its workforce by $\frac{3}{17}$.
   This year the firm will take on 420 people.
   How many people work for the firm at the moment.

   $\frac{3}{17}$ is …… so $\frac{……}{……}$ is …… Number of people now is …… × …… = ……

PRACTICE QUESTIONS

Levels 5–7

1 Complete these calculations.

   a .... × 150 = 54

   b 5.4 ÷ .... = 54

   c 675 × .... = 54   3 marks

2 A cake is divided into four pieces.

   The first three pieces are: $\frac{1}{2}$ of the cake
   $\frac{1}{4}$ of the cake
   $\frac{1}{8}$ of the cake

   a What **fraction** of the cake is the fourth piece?   .... 1 mark

   b Each $\frac{1}{16}$ of the cake has a mass of **20 g**.
   What is the mass of the piece which is $\frac{1}{4}$ of the cake?   .... g   2 marks

3 There are **295** pupils in Year 8 and Year 9 at a school.

	Year 8	Year 9
Boys	60	90
Girls	75	70

   a What **percentage** of these pupils are in **Year 9**?   ....%   2 marks

   b Write the **ratio** of boys to girls in the form 1 : *n*

   Ratio of boys to girls is 1 : ......   2 marks

PRACTICE QUESTIONS

4 Simon keeps $\frac{3}{8}$ of his money in his piggy bank and spends the rest.

    a What **fraction** of his money does he spend? .... 1 mark

    b What **percentage** of his money does he spend? .... % 1 mark

    c He has £24 to start with.
       How much does he **save**? £ .... 1 mark

Level 8

5 The value of a computer is expected to fall by 20% every year.

    a What is the value of a computer which is 2 years old, if it had a value of £800 when new?

       £ .... 2 marks

    b A different computer has a value of £704 when it is 2 years old.
       What was its value when it was new? £ .... 2 marks

    c What single decimal number could you multiply the original value of a computer by, to find its value

       (i) after 2 years? ....

       (ii) after *n* years? .... 2 marks

    d A new computer is advertised with a price of £1500.
       What will its value be after 4 years?

       £ .... 2 marks

    e How many years will it take for the value of any computer to fall by more than half of its original value

       .... years 2 marks

WHAT YOU NEED TO KNOW

3   Estimating

'When I said put brackets on your calculator …'

Levels 3–5

- **BODMAS**
  This tells you which parts of a number question to do first.

First do	**B**rackets
then powers	**O**f
Next do	**D**ivision
and	**M**ultiplication
Then	**A**ddition
and	**S**ubtraction

  $4 + 3 \times 7$
  $= 4 + 21$ BOD**MA**S
  $= 25$ $\times$ before $+$

  $16 \div (2 + 6)$
  $= 16 \div 8$ Bo**D**MAS
  $= 2$ ( ) before $\div$

- **Rounding to the nearest whole number**
  Round 7.6 to the nearest whole number.

  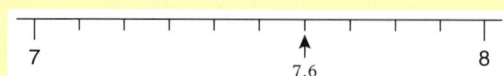

  7.6 is nearer to 8 than to 7.
  It is rounded to 8 to the nearest whole number.

- **Rounding to the nearest 10**
  Round 263 to the nearest 10.

  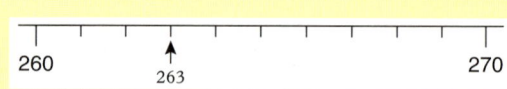

  263 is nearer to 260 than to 270. It is rounded to 260 to the nearest 10.

- **Rounding to the nearest 100**
  Round 750 to the nearest 100.

  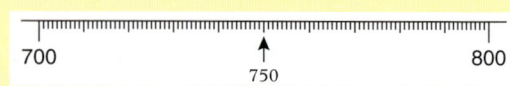

  750 is half way between 700 and 800. It is rounded to 800 to the nearest 100.

WHAT YOU NEED TO KNOW

- **When you use a calculator check that your answer is about the right size.**
  It is very easy to hit the wrong key!

  Work out 3.6 × 239   `3` `.` `6` `×` `2` `3` `9` `=`   Answer = `860.4`

  Estimate: 3.6 is 4 to the nearest whole number.
  239 is 200 to the nearest 100.
  4 × 200 = 800
  800 is near to 860.4 so the answer is probably right.

- **You can also check your answer by writing a new problem.**
  You use the inverse operation.
   226 × 13 = 2938
  To check this you can work out
  either   2938 ÷ 13   or   2938 ÷ 226
           to get 226   to get 13

- **You also need to be able to round decimal numbers.**
  Round 3.14 to 1 decimal place

  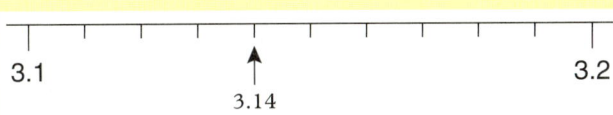

  3.14 is closer to 3.1 than to 3.2
  It is rounded to 3.1 to 1 decimal place (1 dp)

  Round 2.568 to 2 decimal places.

  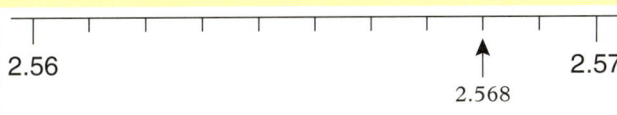

  2.568 is closer to 2.57 than to 2.56
  It is rounded to 2.57 to 2 decimal places (2 dp)

- **You can use this rule of rounding instead of drawing scales.**
  Look at the first unwanted digit. If it is 5, 6, 7, 8 or 9 add one to the last digit you want to keep. If the first 'unwanted' digit is 0, 1, 2, 3, 4, you do not add one.

  5.3**8** is 5.4 to 1 dp   12.57**3**1 is 12.57 to 2 dp.

TEST YOURSELF

Levels 3–5

1 Use BODMAS to work out

   a $15 + 12 \div 4$

     = ... + ...

     = ...

   b $26 - 3 \times 5$

     = ... − ......

     = ...

   c $3 \times 8 - 4$

     = ... − ......

     = ...

   d $7 \times (12 - 8)$

     = ...

     = ...

   e $(8 + 6) \div 2 + 7$

     = ...

     = ... = ...

   f $3^2 + 4 \times 6$

     = ...

     = ...

2   **a** 2.3 = ... to the nearest whole number

    b 46 = ... to the nearest 10

    c 137 = ... to the nearest 10

    d 434 = ... to the nearest 100

    e 99 = ... to the nearest 10

3   **a** 34×176 = ...

    Estimate:

    34 = ... to the nearest 10

    176 = ... to the nearest 100

    ...... × ...... = ...

    b $858 \div 33$ = ...

    Estimate:

    858 = ... to the nearest 100

    33 = ... to the nearest 10

    ...... ÷ ...... = ...

4   **a** 7.85 = ... to 1 dp

    b 6.3174 = ... to 2 dp

    c 13.418 = ... to 1 dp

    d 29.086 = ... to 2 dp

    e $4.857\,31$ = ... to 3 dp

    f 4.6954 = ... to 2 dp

5 A box of 8 tapes costs £11.50.
Find the cost of 1 tape.

   1 tape costs £. . . . . . ÷ . . . . . . = £. . . . . . to the nearest penny

PRACTICE QUESTIONS

Levels 3–5

1 The attendance at a football match was **346 487**.
A local newspaper wants to print the attendance figure.
Write down a **sensible** number for them to use. .... 1 mark

2 Complete the problems.

   a .... ÷ .... = 5   **b** You may use the signs +, −, ×, ÷

                                          24 ...... 3 ...... 1 = 9   2 marks

3 Give two **different** pairs of numbers.

   .... × .... = 72

   .... × .... = 72   2 marks

4 Complete the problems.

   a   3 × .... + ... = 33

   b   (45 − ....) × .... = 90   2 marks

5 Put brackets in the calculations to make the answer given.

   a   3 + 6 + 2 × 4 = 17

   b   6 + 4 + 2 × 3 = 36   2 marks

6 Mark needs 31 litres of paint.
The paint is sold in 5-litre tins.
How many tins of paint should he buy?

   ....   2 marks

7 480 counters are to be put into packets of 50 each for sale.
How many packets will there be to sell?

   ....   2 marks

PRACTICE QUESTIONS

8 The table shows the number of pupils in four neighbouring schools.

School	Number of pupils	To the nearest 100	To the nearest 10
Olney	884		
Mesnes	662		
Heaton	788		
Pendle	906		

4 marks

a Complete the table to show the number of pupils rounded to the nearest **100** pupils, and to the nearest **10** pupils.

b Salus School has **700** pupils, to the nearest **100** pupils, and **650** pupils, to the nearest **10** pupils.
Write down two possible figures for the number of pupils.

……… and ………

2 marks

9 Anita has worked out some questions.
For each question write down another problem she could do to check accurately the answer she has worked out.

a 359 × 13 = 4667   Check: ………………   *1 mark*

b 377 + 761 = 1138   Check: ………………   *1 mark*

c 1830 − 727 = 1103   Check: ………………   *1 mark*

d 5287 ÷ 17 = 311   Check: ………………   *1 mark*

10 Barry has to put 65p into a machine to get a gift.
The machine will take the following coins:

5p, 10p, 20p, 50p   No change is given.

The table shows the number of each coin put in the machine in one day

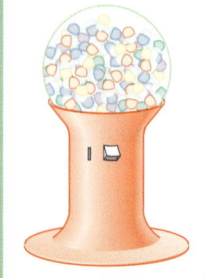

Coin	5 p	10 p	20 p	50 p
Number of coins	21	43	26	27

How many gifts were sold on that day?   ………

3 marks

34

WHAT YOU NEED TO KNOW

Level 7

- **The first significant figure in any number is the first digit which isn't a 0.**
  For most numbers this is the first digit.
  The first significant figure in these numbers is the red digit

  2**5**7   **8**1.5   0.**6**21   0.000**3**16   0.0**2**0 31

- **Rounding to 1 significant figure (1 sf)**
  Look at the digit after the first significant one   0.0**6**4
  Use the rule of rounding.   0.06

- **Be careful to keep the number about the right size.**

  738 is **7**00 to 1 sf.   It is not **7**
  4629 is **5**000 to 1 sf.   It is not **5**

- **When numbers are given to 1 sf they are easy to multiply and divide.**

  400 × **2**000 = **8**00 000   **4** × 100 × **2** × 1000
                                                = **8** × 100 000

  0.0**3** × 0.**2** = 0.00**6**   **3** ÷ 100 × **2** ÷ 10
                                                = **6** ÷ 1000

  0.00**3** × **7**0 = 0.**2**1   **3** ÷ 1000 × **7** × 10
                                                = **21** ÷ 100

  8000 ÷ **2**0 = **4**00 $\dfrac{^4\cancel{8} \times 100\cancel{0}}{\cancel{2} \times \cancel{10}}$ = **4** × 100

  600 ÷ 0.**3** = **2**000   600 ÷ $\dfrac{3}{10}$ = $^2\cancel{6}00 \times \dfrac{10}{\cancel{3}}$

- **Estimating fractions.**
  When you estimate the answer to a question that involves a fraction there is often a better way than just estimating by rounding to 1 sf

  Look at this question. $\dfrac{65.8 \times 52.4}{44}$

  Rounding each number to 1 sf gives $\dfrac{70 \times 50}{40} = \dfrac{3500}{40} = 87.5$

  The actual answer is 78.4 to 3 sf

WHAT YOU NEED TO KNOW

This time start by rounding all the numbers to the nearest whole number.	$\dfrac{66 \times 52}{44} = \dfrac{66 \times 52}{11 \times 4}$
Now you need to spot that the 44 can be split as 11×4.	$= \dfrac{66}{11} \times \dfrac{52}{4}$
This means that you can cancel the fraction like this.	$= 6 \times 13$
78 is a much better approximation than 87.5	$= 78$
Sometimes you have to round so that you can cancel the fraction. Look at this question.	$\dfrac{47.4 \times 79.7}{7.8 \times 9.9}$
Round the numbers in the denominator to the nearest whole number.	$\dfrac{47.4 \times 79.7}{8 \times 9}$
Now look at the numbers in the numerator and round them to multiples of the numbers in the denominator.	$\dfrac{48 \times 81}{8 \times 9} = \dfrac{48}{8} \times \dfrac{81}{9}$
	$= 6 \times 9$
This is close to the actual answer of 53.2 to 3 sf.	$= 54$

Level 8

- **Estimating square roots**
  Round the number to the nearest square number

  $\sqrt{39} \simeq \sqrt{36} = 6$

  If you round $\sqrt{39}$ to 1 sf you get $\sqrt{40}$. You still can't do this in your head.

  The square numbers are 1, 4, 9, 16, 25, ...
  $\quad\quad\quad\quad\quad\quad\quad\quad\quad$ 1×1 2×2 3×3 4×4 5×5

TEST YOURSELF

Level 7

1 Round these to 1 sf.

   a 3.94 = ...
   d 0.007 006 = ...

   b 0.0718 = ...
   e 6503 = ...

   c 26 = ...
   f 9500 = ...

2 **a** 500 × 1000 = ...
   f 0.03 × 5000 = ...

   b 60 × 400 = ...
   g 9000 ÷ 30 = ...

   c 0.003 × 0.1 = ...
   h 1000 ÷ 50 = ...

   d 0.06 × 0.007 = ...
   i 700 ÷ 0.1 = ...

   e 0.2 × 400 = ...
   j 900 ÷ 0.03 = ...

3 Estimate the value of

   a $\dfrac{47.7 \times 35.2}{42}$
   b $\dfrac{28.6 \times 43.8}{4.8 \times 7.1}$

   $= \dfrac{48 \times \ldots}{6 \times \ldots}$
   $= \dfrac{\ldots \times \ldots}{\ldots \times \ldots}$

   $= \dfrac{\ldots}{\ldots} \times \dfrac{\ldots}{\ldots}$
   $= \dfrac{\ldots}{\ldots} \times \dfrac{\ldots}{\ldots}$

   $= \ldots \times \ldots$
   $= \ldots \times \ldots$

   $= \ldots$
   $= \ldots$

Level 8

4 **a** $\sqrt{17} \simeq \sqrt{\ldots} = \ldots$
   b $\sqrt{87} \simeq \sqrt{\ldots} = \ldots$

5 $\sqrt{\dfrac{5^2 \times 62}{2^4}}$

$\sqrt{\dfrac{\ldots \times \ldots}{\ldots}} \simeq \sqrt{\dfrac{\ldots \times \ldots}{\ldots}} = \sqrt{\dfrac{\ldots}{\ldots}} = \sqrt{\ldots} \simeq \sqrt{\ldots} = \ldots$

PRACTICE QUESTIONS

Levels 5–7

1 a A shop sells garden pots for **£2.45**
What is the cost of 4 garden pots?

£ .... 1 mark

b How many garden pots can you buy with **£12**?

.... 1 mark

c The shop also sells the pots in pairs.
One pair of pots costs **£4.49**
How many pairs of pots can you buy with **£12**?

.... pairs 1 mark

d What is the **greatest** number of pots you can buy with **£16**?
You can buy pots either individually or in pairs.

.... pots 1 mark

2 Complete the table below.

Number	Rounded to 1 sf	Rounded to 2 sf	Rounded to 3 sf
0.5182			
10.099			
58.42			
3486			

4 marks

PRACTICE QUESTIONS

3 Mark wants to work out the calculation

$$\frac{81 \times 155}{42.4 \times 2.4}$$

Find an **estimated** answer to this calculation.
Show all your working.

.... 2 marks

Level 8

4 A number has been rounded to 3 significant figures.
The result is the answer 5.12

   a Write down the **smallest** number it could have been.

.... 1 mark

   b Write down the **largest** number it could have been.

.... 1 mark

5 Find an **estimated** answer to the expression:

$$\sqrt{\frac{12.2^3 \times 14.3}{440 \times 9.6^2}}$$

.... 2 marks

WHAT YOU NEED TO KNOW

4 Patterns and sequences

This number pattern is special.
The dots always form a square.

Levels 3–5

- **The square numbers are 1, 4, 9, 16, 25, …**
  You can draw them as square dot patterns.

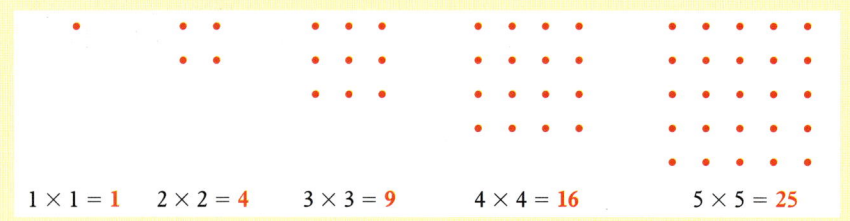

  1 × 1 = 1   2 × 2 = 4   3 × 3 = 9   4 × 4 = 16   5 × 5 = 25

- **The triangle numbers are 1, 3, 6, 10, 15, …**
  You can draw them as triangular dot patterns.

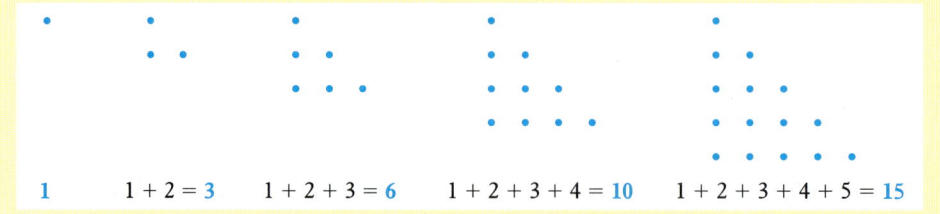

  1   1 + 2 = 3   1 + 2 + 3 = 6   1 + 2 + 3 + 4 = 10   1 + 2 + 3 + 4 + 5 = 15

- **The odd numbers are 1, 3, 5, 7, 9, …**

- **The even numbers are 2, 4, 6, 8, 10, …**
  These are the numbers in the 2 times table.

- The **multiples** of 3 are the 3 times table 3, 6, 9, 12, 15,…
  The **multiples** of 5 are the 5 times table 5, 10, 15, 20, 25, …

WHAT YOU NEED TO KNOW

- **A number that divides exactly into another number is called a factor.**

  To find the factors of 12 look at all the pairs of numbers that multiply to give 12.

  $12 = 1 \times 12$
  $12 = 2 \times 6$
  $12 = 3 \times 4$

  The factors of 12 are **1, 2, 3, 4, 6, 12**

- **Prime numbers have exactly two factors, themselves and 1.**

  $19 = 1 \times 19$ No other two numbers multiply to give 19.
  So 19 is a prime number.

- **The prime numbers are 2, 3, 5, 7, 11, 13, 17, 19, 23, 29, …**

  2 is the first prime number. It is the only even prime number.
  1 is not a prime number.

- **The cube numbers are 1, 8, 27, 64, 125, …**

  $1 \times 1 \times 1 = $ **1**, $2 \times 2 \times 2 = $ **8**, $3 \times 3 \times 3 = $ **27**, $4 \times 4 \times 4 = $ **64**, $5 \times 5 \times 5 = $ **125**, …

- **Square rooting is the opposite of squaring.**

  9 squared $= 9^2 = 9 \times 9 = 81$
  Square root of 81 $= \sqrt{81} = 9$

Level 6

- **A number sequence is a list of numbers that follows a rule.**
  Each number in a sequence is called a term.

- **To find the rule for a sequence**
  Look at how to get from one term to the next

  5,   9,   13,   17,   21, …
  +4 +4   +4   +4

  The rule for this sequence is **add 4**.
  The **2**nd term is $5 + $ **1** lot of **4**
  The **3**rd term is $5 + $ **2** lots of **4**
  The **20**th term is $5 + $ **19** lots of **4**
  The **n**th term is $5 + $ **(n − 1)** lots of **4**

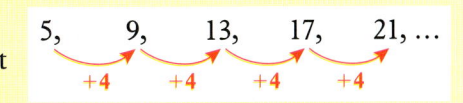

TEST YOURSELF

Levels 3–5

1 Look at these numbers. Write down

   a the multiples of 7 ....

   b the prime numbers ....

   c the triangle numbers ....

   d the cube numbers ....

> 18   13
> 3   10
> 7   1
> 42 21 23
> 2 8 64

2 a Write down the first ten multiples of 4.

....

   b Write down the first ten multiples of 9.

....

   c Write down the number that appears in both lists. ....

3 a 7th square number = ...   **c** 1000 = ...th cube number

   b 6th cube number = ...   **d** 196 = ...th square number

Level 6

4 Look at this sequence   3, 8, 13, 18, 23, ...

  ....

   a The rule is ....

   b The 2nd term is 3 + 1 lot of ....

   c the 50th term is ...... + ...

   d The nth term is ....

PRACTICE QUESTIONS

Levels 3–5

1 Mrs Kay wants **32** desks in her classroom.
She can put them in **4** rows,
with **8** desks in each row.

   a Draw a diagram to show a **different**
way that Mrs Kay can arrange the
32 desks. She must have the same
number of desks in each row.

1 mark

   b Mrs Kay would like to put **6** desks in each row, with the same number of desks in each row.
Explain why Mrs Kay **cannot** arrange the **32** desks in this way.

.... *1 mark*

2 The diagram shows a pattern made out of **black** and **white** tiles.
The diagram is **6** tiles long.

   a Continue the diagram so it becomes **10** tiles long.   *1 mark*

The table shows the details of the tiles used.

Length of pattern (tiles)	1	2	3	4	5	6
Number of white tiles	3	5	8	10	13	15
Number of black tiles	0	1	1	2	2	3

The rule for finding the number of white tiles for a length of an **even** number of tiles is

Number of white tiles = Length ÷ 2 × 5

   b How many **white tiles** are there in a pattern of length **22 tiles**?

.... *1 mark*

   c Complete the rule to show the number of **white tiles** for a length of an **odd** number of tiles:

Number of white tiles = ... *1 mark*

43

PRACTICE QUESTIONS

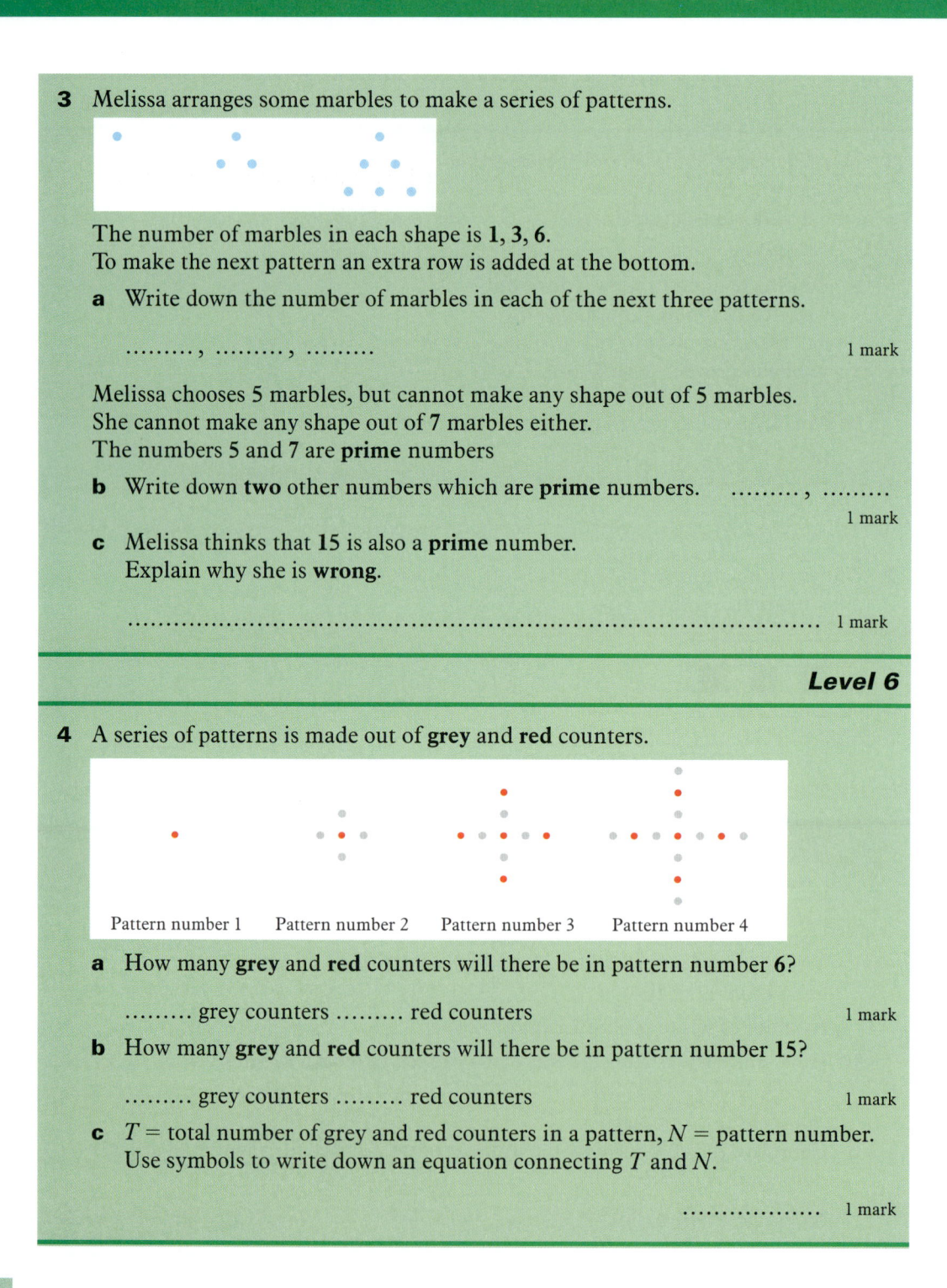

3 Melissa arranges some marbles to make a series of patterns.

The number of marbles in each shape is **1, 3, 6**.
To make the next pattern an extra row is added at the bottom.

a Write down the number of marbles in each of the next three patterns.

.........., ..., ...

1 mark

Melissa chooses 5 marbles, but cannot make any shape out of 5 marbles.
She cannot make any shape out of 7 marbles either.
The numbers 5 and 7 are **prime** numbers

b Write down **two** other numbers which are **prime** numbers.   ...., ...

1 mark

c Melissa thinks that **15** is also a **prime** number.
Explain why she is **wrong**.

.... 1 mark

Level 6

4 A series of patterns is made out of **grey** and **red** counters.

Pattern number 1   Pattern number 2   Pattern number 3   Pattern number 4

a How many **grey** and **red** counters will there be in pattern number **6**?

.... grey counters .... red counters

1 mark

b How many **grey** and **red** counters will there be in pattern number **15**?

.... grey counters .... red counters

1 mark

c T = total number of grey and red counters in a pattern, N = pattern number.
Use symbols to write down an equation connecting T and N.

.... 1 mark

WHAT YOU NEED TO KNOW

Level 7

- **To find a formula for a sequence look at the differences between the terms.**
- **If those differences are the same, find the formula like this.**

    The rule is **+3**.

    You need to **+1** to each multiple of **3** to get the terms of the sequence.
    The formula for the nth term is $3n + 1$.

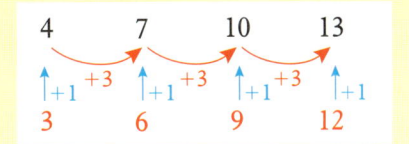

- **If these differences are not the same, look at the second differences.**

    If the second differences are the same the formula for the nth term will contain n^2.
    The number in front of the n^2 is half the second difference.

    The second differences are **+4**.
    The formula will start $2n^2$.

    Write down the sequence for $2n^2$.
    Write down what you need to add to each of these to get the terms of the sequence.
    Use the above method to find the formula for this new sequence.

    The formula is $3n - 1$

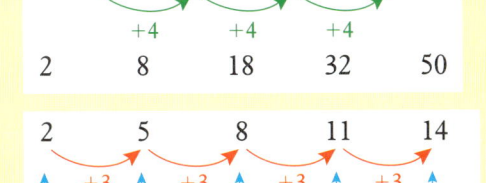

    Put the two parts together to get the formula for the sequence you started with.

    The formula is $2n^2 + 3n - 1$

- **Highest common factor (HCF)**

    The factors of 12 are **1**, **2**, 3, **4**, 6, 12   The common factors of 12 and 20 are 1, 2 and 4
    The factors of 20 are **1**, **2**, **4**, 5, 10, 20   The HCF of 12 and 20 is 4

- **Lowest common multiple (LCM)**

    The multiples of 3 are 3, 6, 9, 12, **15**, 18, 21, 24, 27, **30**, 33 …
    The multiples of 5 are 5, 10, **15**, 20, 25, **30**, 35 …
    The common multiples of 3 and 5 are 15, 30 …
    The LCM of 3 and 5 is 15

TEST YOURSELF

Level 7

1 Find the formula for the
   n^{th} term of this sequence

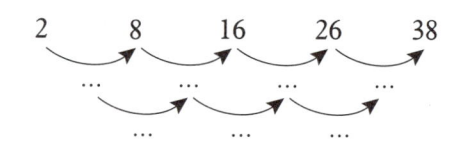

   4   11   18   25   32

   The formula contains ...... n

   The formula is ....

2 The formula for the nth term of a sequence is $4n - 3$.
   Write down

   a the first term $4 \times 1 - 3 = ...$

   b the sixth term $4 \times ... - 3 = ...$

   c the 20th term   .... = ...

3 Find the formula for the
   n^{th} term of this sequence.

   2   8   16   26   38

   The formula will start ....

   To get to the terms of
   the sequence you need   ....

   The formula is   ....

4 The formula for the nth term of a sequence is $3n^2 - n - 2$
   Write down

   a the first term   ....

   b the fifth term   ....

PRACTICE QUESTIONS

Levels 5–7

1 This is a series of patterns made out of grey and blue tiles.

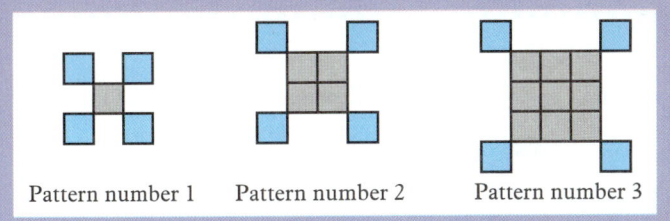

Pattern number 1   Pattern number 2   Pattern number 3

a How many grey and blue tiles will there be in pattern number 7?

.... grey tiles   .... blue tiles   *1 mark*

b How many grey and blue tiles will there be in pattern number 14?

.... grey tiles   .... blue tiles   *1 mark*

c Write expressions to show the total number of grey and blue tiles in pattern number n.

.... grey tiles   .... blue tiles   *1 mark*

d Write an expression to show the total number of tiles in pattern number n.

.... *1 mark*

A different series of patterns is made with the grey and blue tiles.

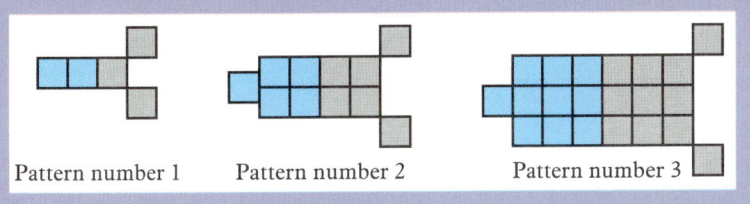

Pattern number 1   Pattern number 2   Pattern number 3

e For this series of patterns write an expression to show the total number of tiles in pattern number n.
Show your working and simplify your expression.

Total number of tiles   .... *2 marks*

WHAT YOU NEED TO KNOW

5 Formulas, expressions and equations

Letters can be used to represent numbers.

Levels 3–5

- **A formula is a set of instructions. It tells you how to work something out.**

- **A formula can be written using words or letters.**

   This formula tells you how to work out the cost, in pounds, of hiring a videotape.

   Cost = 3 × number of nights

   To hire a video for **2** nights

   Cost = 3 × **2** = £6

   This formula tells you how to work out the time, in minutes, needed to roast a turkey.

   Time = 25 + 15 × weight in pounds

   To roast a **12** lb turkey

   Time = 25 + 15 × **12** = 25 + 180 = 205 mins

- **This formula uses letters.**
   The **p**erimeter, p, of a square is given by $p = 4l$
   l is the length of a side of the square.
   Use the formula to find the perimeter of a square with sides of length **8** cm.

   $p = 4 \times 8 = 32$ cm

WHAT YOU NEED TO KNOW

- **You need to be able to write formulas yourself.**

  Pick out the important words and numbers.
  Choose letters to stand for the important words.

  The **cost** of a class trip to a theme park is £**8** for each **pupil** and £**100** for the coach.

  Use C for the cost and p for the number of pupils.

  The formula is $C = 100 + 8p$

- **Algebraic expressions can be simplified.**

  $2a + 3a$ simplifies to $5a$
  $6b - 3b + b$ simplifies to $4b$

  Remember that b means $1b$.

Level 6

- **Collecting like terms**

  $5p + 6q - 2p + 5q = 3p + 11q$
  $7x - 2 - 9x + 6 = -2x + 4$

- **Multiplying out a bracket**

  Multiply each term in the bracket by the term outside the bracket.

  $4(x + 2) = 4x + 8$ $5g(g + 5) = 5g^2 + 25g$

- **Factorising**

  This is the reverse of multiplying out a bracket.

$12p - 18 = 6(2p - 3)$	6 is the highest common factor of 12 and 18.
$3x^2 + 2x = x(3x + 2)$	This time you take an x out.
$d^3 + 5d^2 - 12d = d(d^2 + 5d - 12)$	Make sure you have the same number of terms inside the bracket as there were in the question.

49

WHAT YOU NEED TO KNOW

- **You can use algebra to solve equations.**
  You need to get the letter by itself on one side of the equation. Look at what is being done to the letter. Do the opposite of this to both sides of the equation.

$x + 4 = 7$ take 4 from both sides $x + 4 - 4 = 7 - 4$ $x = 3$	$p + 8 = 2$ take 8 from both sides $p + 8 - 8 = 2 - 8$ $p = -6$
$5r = 35$ divide both sides by 5 $\dfrac{5r}{5} = \dfrac{35}{5}$ $r = 7$	$\dfrac{y}{6} = 4$ multiply both sides by 6 $\dfrac{y}{6} \times 6 = 4 \times 6$ $y = 24$

- **Some equations are more difficult.** More than one step is needed.

$2x + 7 = 16$ First, take 7 from both sides $2x + 7 - 7 = 16 - 7$ $2x = 9$ Now, divide both sides by 2 $\dfrac{2x}{2} = \dfrac{9}{2}$ $x = 4\frac{1}{2}$	$7x + 4 = 5x + 10$ This equation has xs on both sides First, take $5x$ from both sides $7x - 5x + 4 = 5x - 5x + 10$ $2x + 4 = 10$ Now take 4 from both sides $2x + 4 - 4 = 10 - 4$ $2x = 6$ divide both sides by 2 $x = 3$

- **When you have a bracket in an equation, multiply the bracket out first.**

  Multiply out the bracket.
  $$2(x + 7) = 15$$
  $$2x + 14 = 15$$
  $$2x = 1 \quad \text{so } x = \tfrac{1}{2}$$

TEST YOURSELF

Levels 3–5

1 a A plumber charges a call out charge of £25.
He then charges £15 an **hour**.
Write down the formula for the total **cost**.

....

b Find the cost if the plumber is working for 4 hours.

....

2 Write down a formula for the perimeter, P, of this rectangle.

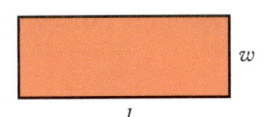

w

l

....

Level 6

Simplify
3 a $6c + 5d - c + d$
= ...... + ...

Factorise
c $9p - 12$
= ...... (...... − ......)

Multiply out the bracket
b $5(2q + 7)$
= ...... + ...

Factorise
d $z^3 + 4z^2 - 8z$
= ...... (...... + ... − ......)

4 Solve these equations

a $x + 7 = 13$

$x + 7 \;\text{......} = 13 \;\text{......}$

$x = \text{......}$

c $5x - 28 = x - 6$

$5x \;\text{......} - 28 = x \;\text{......} - 6$

$\text{......} - 28 = -6$

$\text{......} = \text{............}$

$x = \text{......}$

b $\dfrac{x}{6} - 3 = 4$

$\dfrac{x}{6} - 3 \;\text{......} = 4 \;\text{......}$

$\dfrac{x}{6} = \text{......}$

$\dfrac{x}{6} \times \text{......} = \text{......} \times \text{......}$

$x = \text{......}$

d $4(x - 3) = 28$

$\text{......} = \text{......}$

$4x = \text{......}$

$x = \text{......}$

51

PRACTICE QUESTIONS

Levels 3–5

1 Here are some expressions:

$x \div 2 \quad x + 2 \quad x^3 \quad x + x \quad x^2 \quad 3x \quad x \times 2 \quad x \div 3 \quad x \div x \quad 2 + x$

a Write down the expression that is the same as $x \times x$

.... 1 mark

b Write down the expression that is the same as $\dfrac{x}{3}$

.... 1 mark

c **Two** expressions are the same as $2x$.
Write down these two expressions.

.... 1 mark

d Write down a **new** expression which will always be the same as $3x + x$.

.... 1 mark

2 There are *n* counters in a bag.

a Two of these bags, each with *n* counters, are put together to make a new bag.
Write an expression for the number of counters that are in the new bag.

.... 1 mark

b **Five** counters are then removed from the new bag.
Write an expression for the number of counters that are now in the new bag.

.... 1 mark

3 This **square** tile has edges of length *n* centimetres.
Six of the square tiles are put together to make a shape.

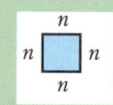

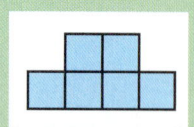

a Write an expression for the **perimeter** of the shape.
Simplify your expression.

.... cm   2 marks

PRACTICE QUESTIONS

b The perimeter of the shape is **60** cm.
Use your answer to part **a** to write an equation using **n**.
Solve your equation to find the **value of n**.

$n =$ ....   2 marks

Level 6

4 Sapna had **3m** meal vouchers, but has used **4** of them.
Dale had **2m** meal vouchers, has used none, but has gained **one extra** voucher.

   a Write down **expressions**, in *m*, for the number of meal vouchers that Sapna and Dale each have.

Sapna ....   Dale ....   2 marks

   b Sapna and Dale now have the **same** number of meal vouchers.
Write down an equation to show this.

....   1 mark

   c Solve the equation to find **m**.

$m =$ ....   1 mark

5 $y = x^2 + x - 3$

Find a value of *x*, to 1 dp, that gives the value of *y* closest to 0.
Two values for *x* have already been worked out.

x	1	2					
y	−1	3					

$x =$ ....

WHAT YOU NEED TO KNOW

Level 7

- **The signs < ≤ > ≥ are all called inequality signs.**
  < means less than > means greater than
  ≤ means less than or equal to ≥ means greater than or equal to

- **You can show inequalities on a number line.**

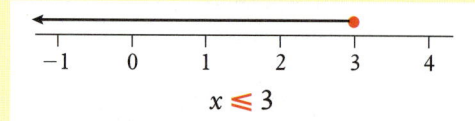

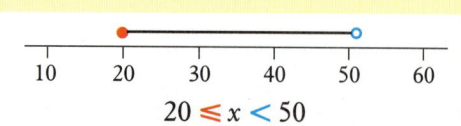

  $x \leq 3$

  $20 \leq x < 50$

- **You solve inequalities like equations.**

  $4x - 6 \geq 11$

  $4x - 6 + 6 \geq 11 + 6$

  $4x \geq 17$

  $x \geq \frac{17}{4}$

  $\frac{x}{3} + 4 < 2$

  $\frac{x}{3} + 4 - 4 < 2 - 4$

  $\frac{x}{3} < -2$

  $x < -6$

- **Sometimes you have to give possible values.**
  If $-4 \leq x < 3$ and x is an integer
  then x can take the values $-4, -3, -2, -1, 0, 1, 2$.
  An integer is a whole number.

- **Changing the subject of a formula**
  This works like solving equations.
  Get the new subject letter by itself.
  Write the new formula with the subject letter on the left hand side.

  Make s the subject.

  $p = 3s - r$

  $p + r = 3s$

  $\frac{p + r}{3} = s$

  $s = \frac{p + r}{3}$

  Make b the subject.

  $a = \sqrt{b + 3c}$

  $a^2 = b + 3c$

  $a^2 - 3c = b$

  $b = a^2 - 3c$

WHAT YOU NEED TO KNOW

Level 8

- **Inequalities can have two variables, usually x and y.**

    These can be shown using a graph
    The red line is $y = 2x + 1$
    The red shading shows $y \geq 2x + 1$

    The blue line is $y = x - 2$
    The blue shading shows $y < x - 2$
    The line is dashed to show that
    it is not included.

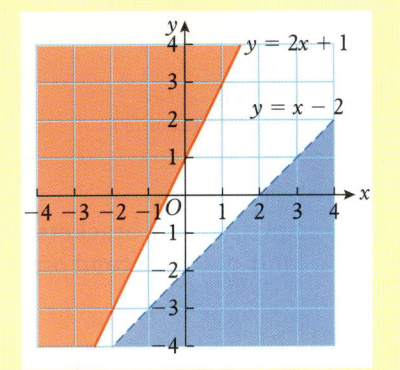

- **A region can be described using more than one inequality.**

    Show the region given by

    $-2 < x \leq 3 \qquad y < 1 \qquad y \geq x - 2$

    You can shade the parts that you don't want.
    The answer is the region, R, which is not shaded.

    Always make it clear on your diagram
    which is the required region.

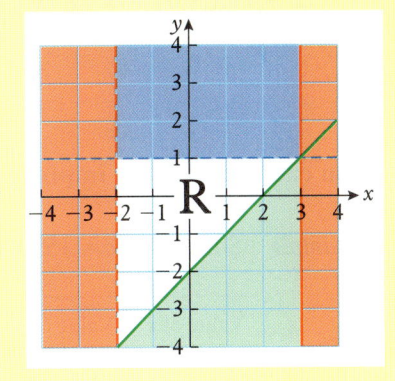

- **Substituting into formulas**

    You need to be able to deal with fractions, decimals and negative numbers.
    If $y = 4x^2 - 3$ find y when **a** $x = \frac{3}{4}$ **b** $x = 2.7$ **c** $x = -5$

    a $y = 4 \times (\frac{3}{4})^2 - 3$ **b** $y = 4 \times (2.7)^2 - 3$ **c** $y = 4 \times (-5)^2 - 3$
    $\quad = 4 \times \frac{9}{16} - 3$ $= 4 \times 7.29 - 3$ $= 4 \times 25 - 3$
    $\quad = -\frac{3}{4}$ $= 26.16$ $= 97$

- **Solving complicated equations**

    $\dfrac{17 - x}{4} = 2 - x$ Multiply by the 4 to get rid of the fraction.

    $17 - x = 4(2 - x)$ Now solve in the usual way.

    $17 - x = 8 - 4x$ So $3x = -9$ and $x = -3$

WHAT YOU NEED TO KNOW

- **Brackets and powers**
  $x^3(4x^2 - 7) = 4x^5 - 7x^3$ 　　　　　 $2a^2b(a^2 + 3b) = 2a^4b + 6a^2b^2$

- **Factorising**
  $6t^4 + 9t^3 = 3t^3(2t + 3)$ 　　　　　 $12a^3 - 4a^2 = 4a^2(3a - 1)$

- **Multiplying out 2 brackets**
  You multiply the second bracket by each term in the first bracket.

  $(s + 3)(2s - 5) = s(2s - 5) + 3(2s - 5)$ 　　　 $(x + 4)^2 = (x + 4)(x + 4)$
  $\qquad\qquad\quad = 2s^2 - 5s + 6s - 15$ 　　　　　　　　　 $= x^2 + 4x + 4x + 16$
  $\qquad\qquad\quad = 2s^2 + s - 15$ 　　　　　　　　　　　　 $= x^2 + 8x + 16$

  $(x - 7)(x + 7) = x^2 + 7x - 7x - 49$ 　　　　　 $(2a + b)^2 = (2a + b)(2a + b)$
  $\qquad\qquad\quad = x^2 - 49$ 　　　　　　　　　　　　　　　 $= 4a^2 + 2ab + 2ab + b^2$
  $\qquad\qquad\qquad\qquad\qquad\qquad\qquad\qquad\qquad\quad = 4a^2 + 4ab + b^2$

- **You can use trial and improvement to solve more complicated equations.**

  Example　Solve $x^3 + x = 1590$ giving your answer to 1 dp.

Value of x	Value of $x^3 + x$	
11	1342	too small
12	1740	too big
11.5	1532.375	too small
11.6	1572.496	too small
11.7	1613.313	too big
11.65	1592.817 125	too big

  x is between **11** and **12**
  x is between **11.5** and **12**
  x is between **11.6** and **12**
  x is between **11.6** and **11.7**
  x is between **11.6** and **11.65**

  11.6　　11.65　　11.7

  x must be somewhere in the green part of the number line. Any number in the green part rounds down to 11.6 to 1 dp.

  Answer:　$x = 11.6$ to 1 dp.

TEST YOURSELF

Level 7

1 Solve these inequalities and show your answers on number lines.

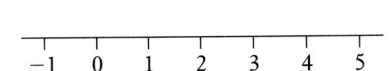

    a $2x + 3 \leq 11$ **b** $\dfrac{x}{5} - 2 > 1$

        $2x \leq \ldots\ldots$ $\ldots\ldots > \ldots\ldots$

        $\ldots\ldots\ldots$ $\ldots\ldots\ldots$

2 If $-1 < x \leq 4$ and x is an integer then x can take the values $\ldots\ldots\ldots\ldots\ldots\ldots\ldots$

3 Make r the subject of

    a $x = 2r - c^2$ **b** $p = (r - s)^2$

        $\ldots\ldots = 2r$ $\ldots\ldots = (r - s)$

        $\ldots\ldots\ldots\ldots$ $\ldots\ldots\ldots\ldots$

        $r = \ldots\ldots\ldots$ $\ldots\ldots\ldots\ldots$

4 Factorise

    a $14q - 8$ **b** $16a^3 - 8a$

        $\ldots\ldots\ldots\ldots$ $\ldots\ldots\ldots\ldots$

Level 8

5 Write down the inequalities shown by the shading.

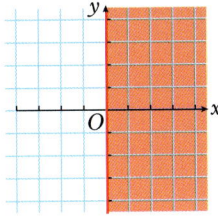

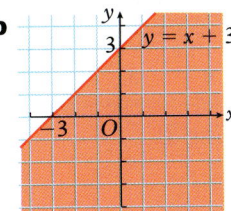

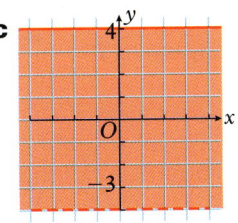

    a   **b** $y = x + 3$ **c**

6 Multiply out

    a $(x + 4)(x - 3)$ **b** $(2x - 3)(x - 6)$

        $\ldots\ldots\ldots\ldots$ $\ldots\ldots\ldots\ldots$

        $\ldots\ldots\ldots\ldots$ $\ldots\ldots\ldots\ldots$

PRACTICE QUESTIONS

Levels 5–7

1 The table shows the result of adding pairs of expressions. Complete the table.

A	B	A + B
x	y	
$3a$	$5a$	
$2p$		$2p - 5q$
$5x + 2y$	$3y - 4w$	
	$6k$	$2k$
x^2	$3x^2$	

3 marks

2 A weatherman reported, 'The temperature this summer never quite reached 24 degrees.'
   a Shade the temperature scale to show the **maximum** temperature.
   1 mark
   b T = temperature
   Write down an inequality, using T, to show the statement of the weatherman.

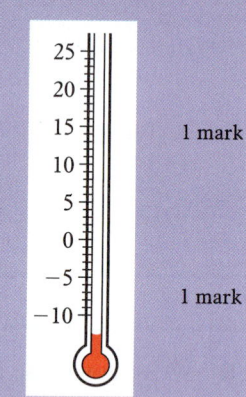

   ....
   1 mark

   c The temperature for the winter months could be represented by the diagram

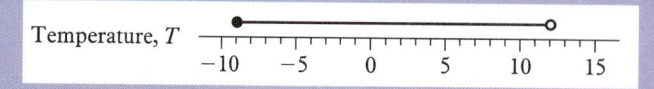

   Temperature, T $\quad -10 \quad -5 \quad 0 \quad 5 \quad 10 \quad 15$

   Write down an inequality, using T, to show this range of temperature.

   ....
   1 mark

3 Find the values of P and Q when $x = 5$ and $y = 3$

   a $P = \dfrac{x^2(x - 1)}{4}$

   $P = $ ....

   b $Q = \dfrac{5y^3}{6}$

   $Q = $ ....

   2 marks

PRACTICE QUESTIONS

4 The diagram shows a **square**.
The expressions show the length of a side of the square, in centimetres.

Calculate the value of x, and use it to find the area of the square.

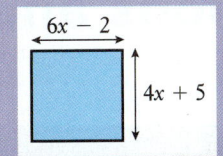

.... cm²   2 marks

Level 8

5 The diagram shows some straight line graphs.

  a Write down **three** inequalities which can be used to describe the shaded region.   3 marks

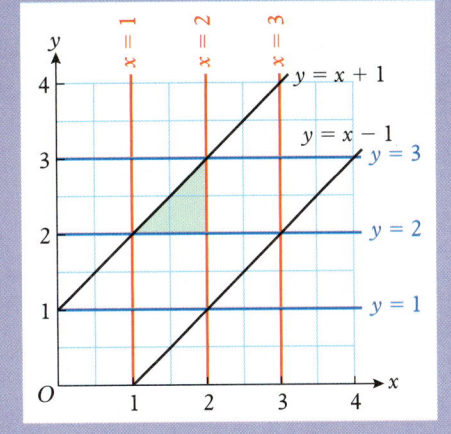

  b On the diagram shade the region which is described by the inequalities

$y \geqslant 1$ $y \leqslant x - 1$ $x \leqslant 3$ 2 marks

6 Multiply out and simplify these expressions:

  a $2(x - 3) - (x - 7)$ **b** $(x - 2)(x + 3)$ **c** $(x + 5)^2$ 3 marks

7 Solve these equations.

  a $5 - 3x = 6x + 11$ **b** $2(x - 1) = 6$ **c** $\dfrac{3x}{x - 1} = 5$

$x = $ .... $x = $ .... $x = $ ....   6 marks

WHAT YOU NEED TO KNOW

6 Functions and graphs

You use co-ordinates to find the position of a point.

Levels 3–5

- **You need two co-ordinates to describe the position of a point.**

  The co-ordinates are
  Sun   (0, 0)
  Mercury (−1, 1)
  Venus   (2, −1)
  Earth   (−2, −2)

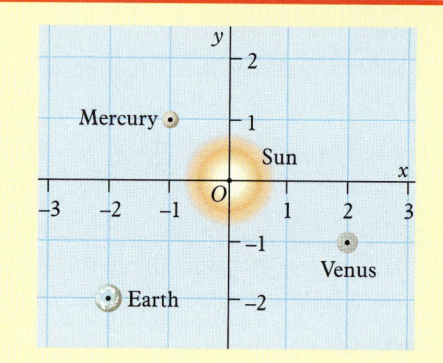

Level 6

- **To draw a graph**
  (1) Make a table of values.
  (2) Plot the points and join them up.
  Draw the graph of $y = 2x + 1$

x	−1	0	1	2
y	−1	1	3	5

  $2 \times (-1) + 1$ ↗   ↖ $2 \times (1) + 1$

  The line $y = 2x + 1$ carries on forever in both directions. What you have drawn is part of the line. It is called a line segment.

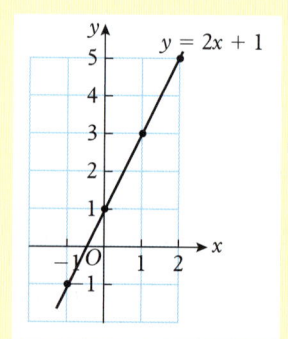

WHAT YOU NEED TO KNOW

- **The equation of any straight line can be written as $y = mx + c$.**

    m is the gradient of the line. It tells you how steep the line is. c is where the line crosses the y axis.

    The red line has gradient 1 and crosses the y axis at 3.

    The blue line has gradient -2 and crosses the y axis at 1.

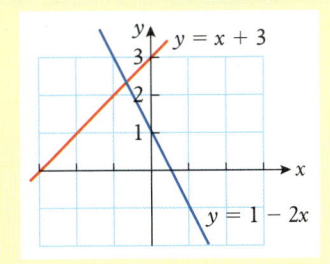

- **Vertical lines have equations starting $x =$**

- **Horizontal lines have equations starting $y =$**

- **Parallel lines have the same gradient.**

- **Checking to see if a point lies on a line.**
    Does the point (2, 5) lie on the line $y = 3x - 2$?
    When $x = 2$, $y = 3 \times 2 - 2 = 4$
    so (2, 4) lies on the line.
    (2, 5) does not lie on the line.

- **Finding missing co-ordinates**
    The point (5, a) lies on the line $y = 4x - 1$. Find a.

    When $x = 5$, $y = 4 \times 5 - 1 = 19$
    so $a = 19$

    The line $y = x + 2$ crosses the line $y = 1$ at P. Find the co-ordinates of P.

    When $y = 1$, $1 = x + 2$ so $x = -1$,
    The co-ordinates of P are $(-1, 1)$.

- **Some equations do not have y on its own.**
    To draw the graph of $3x + 4y = 12$
    find where the graph crosses each axis

    When $x = 0$ When $y = 0$
       $4y = 12$ $3x = 12$
        $y = 3$ $x = 4$
    This gives (0, 3)   This gives (4, 0)

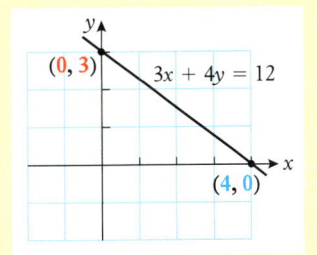

61

TEST YOURSELF

Levels 3–5

1 Fill in the co-ordinates.

A (......, ...)   B (......, ...)   C (......, ...)

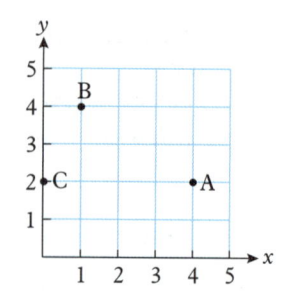

2 Plot these points
D (1, 3)   E (4, 5)   F (3, 0)

Level 6

3 Write down the equation of the red line.

....

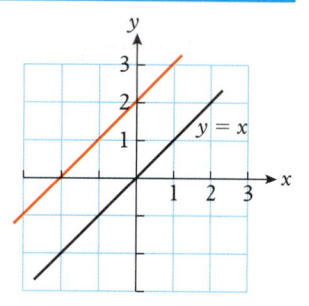

4 Write down the equation of each of these lines.
Choose from
$y = -x$, $y = 2x + 2$, $x = 2$, $y = -x - 3$

A ....   C ....

B ....   D ....

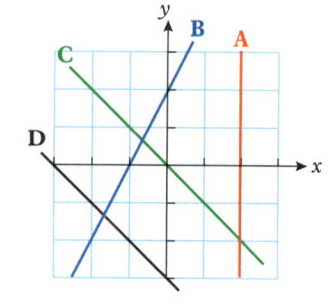

5 Draw the graph of $3x - 2y = 6$

When $x = 0$ When $y = 0$

.... = ... = ...

.... = ... = ...

This gives (......, ...)   This gives (......, ...)

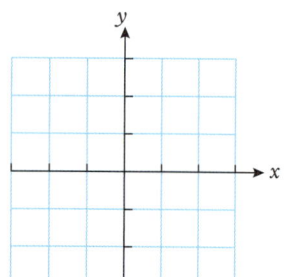

PRACTICE QUESTIONS

Levels 3–5

1 These triangles make a pattern on a grid.
Each triangle is numbered.
The **bottom right hand corner** of each triangle is marked with a letter.

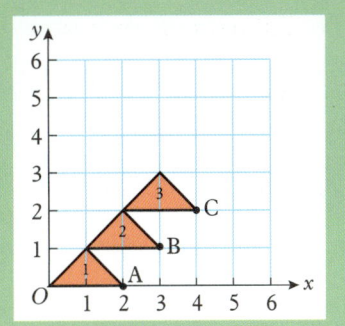

a Write down the co-ordinates of the **corner** marked

A (,)   B (,)   C (,)

2 marks

b What are the co-ordinates of the bottom right hand corner of triangle **number 10**?

(,)

Explain how you worked out your answer.

2 marks

c Explain why **(10, 9) cannot** be a **corner** of a triangle.

1 mark

2 a Plot each of the points P, Q, R and S on the grid. Join the points together with straight lines.

P (0, 0)   R (2, 2)
Q (0, 2)   S (2, 0)

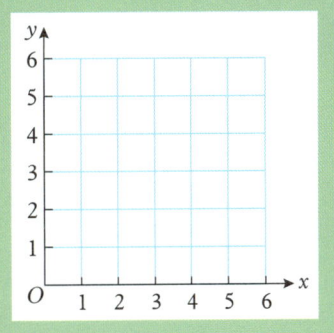

2 marks

b Give the name of the **shape** that you have drawn.

.... *1 mark*

PRACTICE QUESTIONS

  c Multiply each of the co-ordinates P, Q, R and S **by 3**, and plot these points on the same axes

      P (0, 0) × 3 → (,)

      Q (0, 2) × 3 → (,)

      R (2, 2) × 3 → (,)

      S (2, 0) × 3 → (,)

                                                                                                                            2 marks

  d How many **times larger** is the **area** of this shape, compared with the area of the original shape?

                                                                                            .... 1 mark

Level 6

3 The diagram shows the graph of the straight line $y = x + 1$.

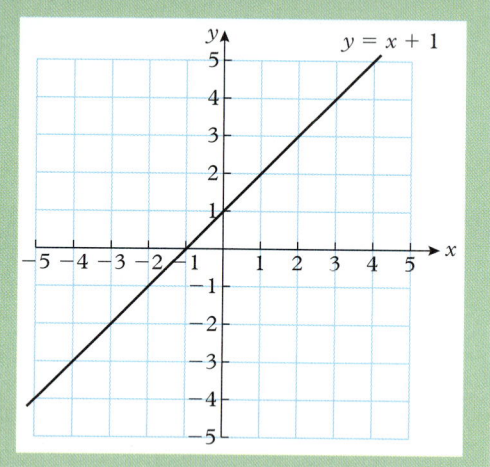

  a Draw the graph of the straight line $y = x + 2$.
Label your graph.   1 mark

  b Draw the graph of the straight line $y = 2x + 2$.
Label your graph.   1 mark

  c Write down the equation of any straight line which goes through (0, 0).

                                                                   $y = $ .... 1 mark

  d Write down the equation of a straight line which is parallel to $y = x + 1$, and goes through $(-2, 5)$.

                                                                   $y = $ .... 1 mark

WHAT YOU NEED TO KNOW

Level 7

- **The point where two lines cross is called the point of intersection.**
  These lines cross at the point (2, 3).

- $x = 2$, $y = 3$ is the solution to the simultaneous equations

  $$y = 3x - 3$$
  $$x + y = 5$$

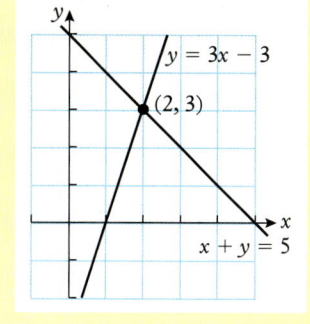

- **Mid-point of a line**

  To find the mid-point of a line, add the co-ordinates of the end points and divide by 2.

  Mid-point of AB is $\left(\dfrac{1 + 5}{2}, \dfrac{3 + 13}{2}\right)$

  $= (3, 8)$

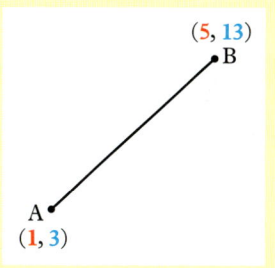

- **To solve a pair of simultaneous equations**

  either   draw the graphs of the 2 equations and find where they cross
  or   use algebra.

- **Solving simultaneous equations using algebra**

  Solve (1) $5x + y = 20$
        (2) $2x + y = 11$

  Subtract the equations to get rid of y

  $$3x = 9$$
  $$x = 3$$

  Put $x = 3$ into (1) to find y

  $$5 \times 3 + y = 20$$
  $$y = 5$$

  The solution is $x = 3, y = 5$

  Use (2) to check your answer

  $$2x + y = 2 \times 3 + 5$$
  $$= 11 \checkmark$$

  Solve (1) $3x - y = 19$
        (2) $x + y = 1$

  Add the equations to get rid of y

  $$4x = 20$$
  $$x = 5$$

  Put $x = 5$ into (1) to find y

  $$3 \times 5 - y = 19$$
  $$y = -4$$

  The solution is $x = 5, y = -4$

  Use (2) to check your answer

  $$x + y = 5 - 4$$
  $$= 1 \checkmark$$

WHAT YOU NEED TO KNOW

- **You sometimes need to multiply one or both of the equations before you add or subtract.**

Solve this pair of simultaneous equations $\quad 2x + 6y = 13$
$\qquad\qquad\qquad\qquad\qquad\qquad\qquad\qquad\qquad 4x - 2y = 5$

Number the equations $\qquad\qquad$ (1) $\quad 2x + 6y = 13$
$\qquad\qquad\qquad\qquad\qquad\qquad$ (2) $\quad 4x - 2y = 5$

You need to multiply equation (2) by **3**
so that you have $6y$ in $\qquad\qquad\qquad\qquad\qquad 2x + 6y = 13$
each equation $\qquad\qquad\qquad$ (2) × **3** $\quad \underline{12x - 6y = 15}$

Add to get rid of y $\qquad\qquad\qquad\qquad\quad 14x \qquad\;\; = 28$
This finds x $\qquad\qquad\qquad\qquad\qquad\qquad\qquad\quad x = 2$

Use equation (1) to find y $\qquad$ Put $x = 2$ in equation (1)
$\qquad\qquad\qquad\qquad\qquad\qquad\qquad 2 \times 2 + 6y = 13$
$\qquad\qquad\qquad\qquad\qquad\qquad\qquad\quad 4 + 6y = 13$
$\qquad\qquad\qquad\qquad\qquad\qquad\qquad\qquad\; 6y = 9$
$\qquad\qquad\qquad\qquad\qquad\qquad\qquad\qquad\quad y = 1.5$
$\qquad\qquad\qquad\qquad$ The answer is $x = 2$, $y = 1.5$

Use equation (2) to $\qquad$ Check $\quad 4x - 2y = 4 \times 2 - 3$
check your answer $\qquad\qquad\qquad\qquad\qquad\;\; = 5$ ✓

Solve this pair of simultaneous equations $\quad 3x + 5y = 30$
$\qquad\qquad\qquad\qquad\qquad\qquad\qquad\qquad\qquad 2x + 3y = 19$

Number the equations $\qquad\qquad$ (1) $\quad 3x + 5y = 30$
$\qquad\qquad\qquad\qquad\qquad\qquad$ (2) $\quad 2x + 3y = 19$

Multiply equation (1) by 2 $\qquad\qquad\quad 6x + 10y = 60$
Multiply equation (2) by 3 $\qquad\qquad\quad \underline{6x + \;\;9y = 57}$

You can now subtract to get rid of x $\qquad\qquad y = 3$
Put $y = 3$ in equation (1) $\qquad\qquad\qquad 3x + 15 = 30$
$\qquad\qquad\qquad\qquad\qquad\qquad\qquad\qquad 3x = 15$
$\qquad\qquad\qquad\qquad\qquad\qquad\qquad\qquad\; x = 5$
$\qquad\qquad\qquad\qquad$ The answer is $x = 5$, $y = 3$

Check using equation (2) $\qquad$ Check $\quad 2x + 3y = 2 \times 5 + 3 \times 3$
$\qquad\qquad\qquad\qquad\qquad\qquad\qquad\qquad\qquad = 19$ ✓

WHAT YOU NEED TO KNOW

Level 8

- **You need to be able to recognise these graphs.**

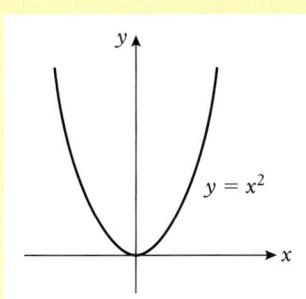

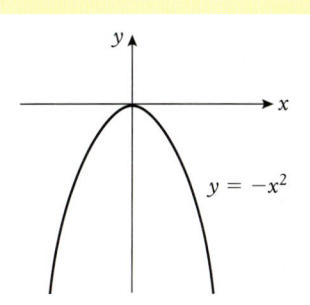

  $y = x^2$

  $y = -x^2$

  Any quadratic graph will look like ∪ or ∩
  An equation for a quadratic graph will have an x^2 term but no higher powers of x.

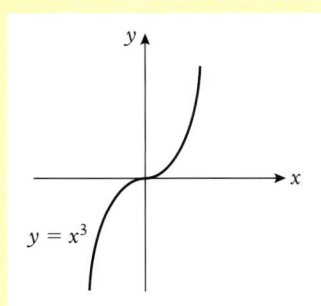

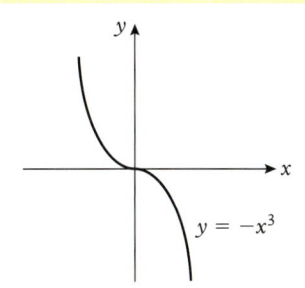

  $y = x^3$

  $y = -x^3$

  Other cubic graphs will look like one of these or ∿ or ∾
  An equation for a cubic graph will have an x^3 term but no higher powers of x.

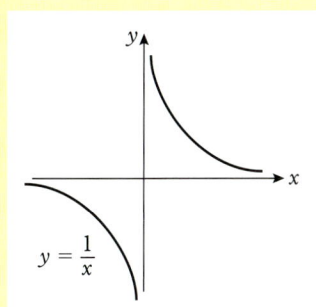

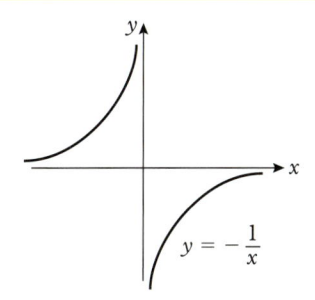

  $y = \dfrac{1}{x}$

  $y = -\dfrac{1}{x}$

  All reciprocal graphs will look like one of these.
  Reciprocal graphs always have two parts.

WHAT YOU NEED TO KNOW

- **Look for these features when describing or sketching graphs that model real situations,** e.g. filling containers with liquid, speed around a track.

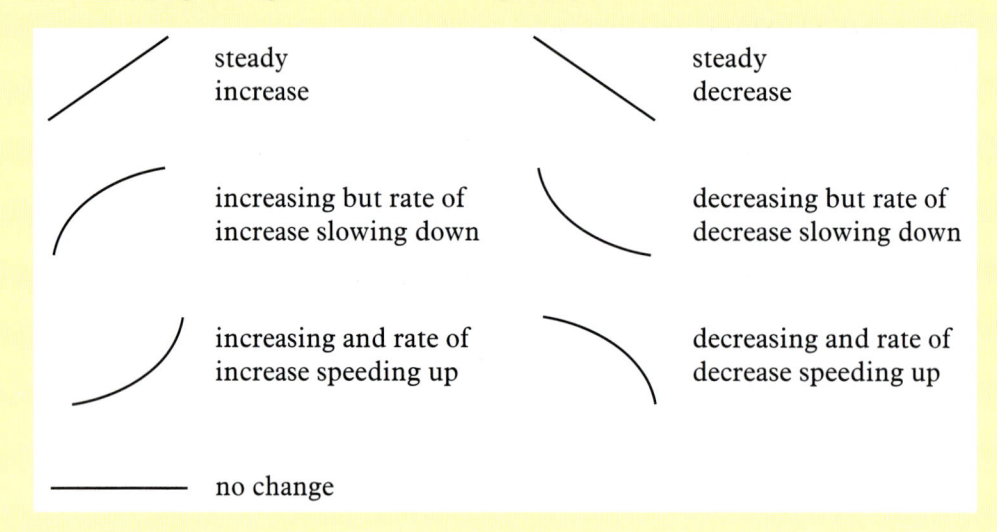

  steady increase

  steady decrease

  increasing but rate of increase slowing down

  decreasing but rate of decrease slowing down

  increasing and rate of increase speeding up

  decreasing and rate of decrease speeding up

  no change

- You can describe a graph using these features.

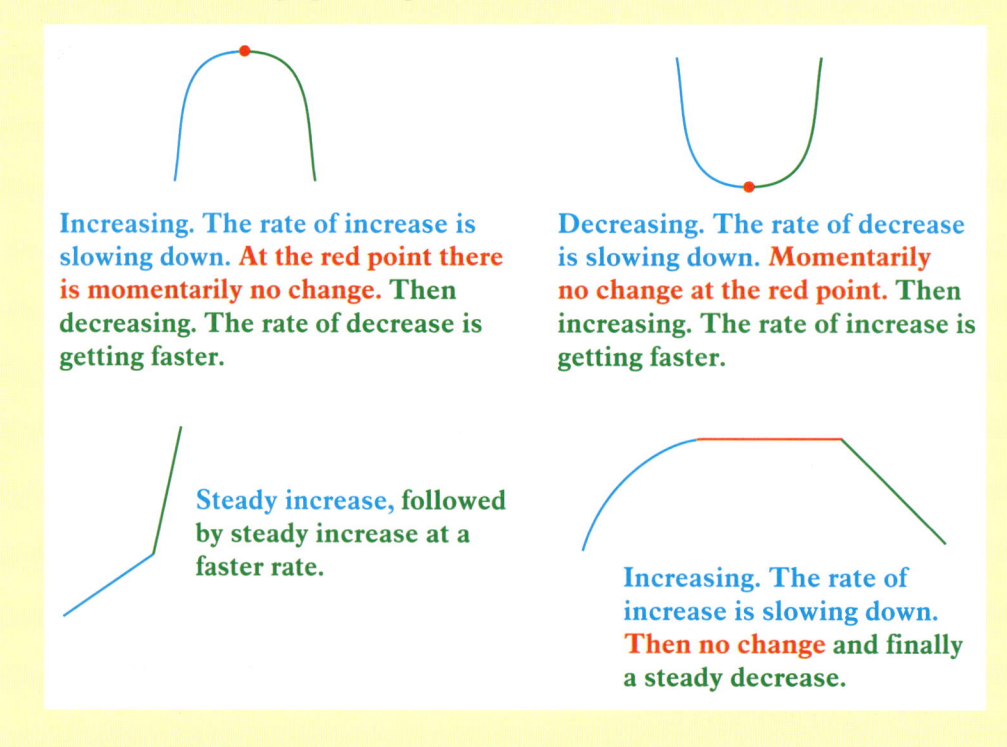

  Increasing. The rate of increase is slowing down. At the red point there is momentarily no change. Then decreasing. The rate of decrease is getting faster.

  Decreasing. The rate of decrease is slowing down. Momentarily no change at the red point. Then increasing. The rate of increase is getting faster.

  Steady increase, followed by steady increase at a faster rate.

  Increasing. The rate of increase is slowing down. Then no change and finally a steady decrease.

TEST YOURSELF

Level 7

1 Solve these pairs of simultaneous equations:

a (1) $2x + y = 14$
  (2) $3x - y = 6$

Adding   …… = ……

  x = ……

Put x = …… into (1)

  $2 \times$ …… $+ y$ = ……

  y = ……

the solution is x = ……, y = ……

Check in (2)

  $3 \times$ …… $-$ …… $=$ ……

b (1) $4x + 2y = 10$
  (2) $5x + 3y = 12$

Multiply (1) by …… and (2) by ……

  …… = ……
  …… = ……

Subtracting …… = ……
  …… = ……

Put …… = …… into (1)

  …… = ……
  …… = ……

the solution is ……………

Check in (2) ………………

Level 8

2 Write down the letter of the graph by its equation.

A   B 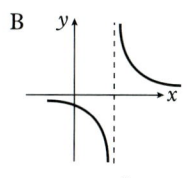   C 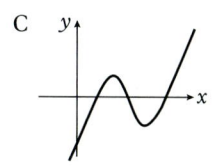   D

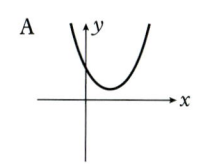

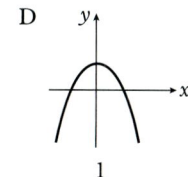

$y = x^3 - 6x^2 + 11x - 6$ $y = 4 - x^2$ $y = x^2 - 4x + 7$ $y = \dfrac{1}{x-3}$

Graph ……   Graph ……   Graph ……   Graph ……

3 This flask is filled with water at a constant rate.
Sketch the graph to show how the height of the water changes and explain the features of your graph.

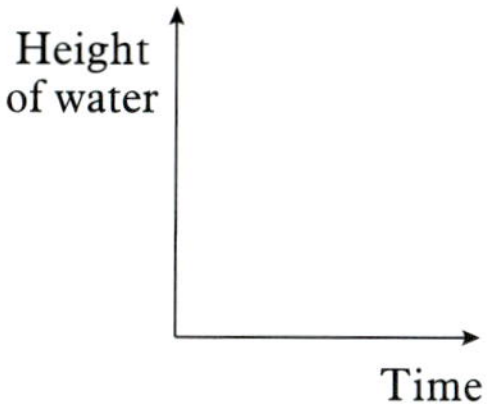

Height of water / Time

……………………………………………
……………………………………………
……………………………………………
……………………………………………

69

PRACTICE QUESTIONS

Levels 5–7

1 a Write down the **equation** of the line through A and B ....

Write down the **equation** of the line through B and C ....

2 marks

b Fill in the gaps below:

$y - x = 2$ is the equation of the line

through .... and .... 1 mark

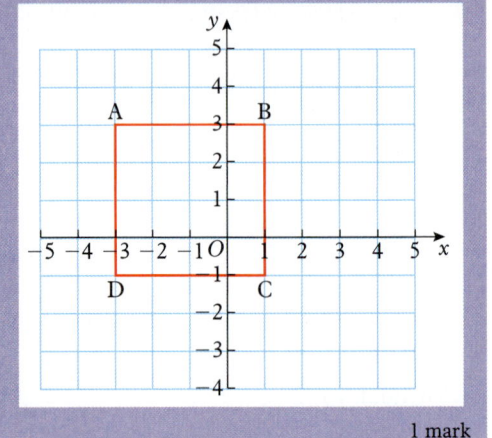

c One of the lines of symmetry of the square is $x = -1$.
On the diagram, **draw and label** the line $x = -1$. 1 mark

Write down the equation of **one other** line of symmetry.

.... 1 mark

d The line $y - x = 2$ crosses the line $2y + x = 22$

Solve the simultaneous equations (1) $2y + x = 22$
Show your working.   (2) $y - x = 2$

$x = $ .... $y = $ ....   2 marks

e Write down the co-ordinates of the point where the line $y - x = 2$ crosses the line $2y + x = 22$.

(......, ...) 1 mark

2 Solve these simultaneous equations to find the value of x and y.

  (1) $8x + 6y = 10$
  (2) $30x - 6y + 1 = 10$

Show your working.

$x = $ .... $y = $ ....   3 marks

PRACTICE QUESTIONS

3 The diagram shows the graphs of the equations

$$4y = x + 2$$
$$4y = -x + 6$$

Use the diagram to write down the solutions to these simultaneous equations.

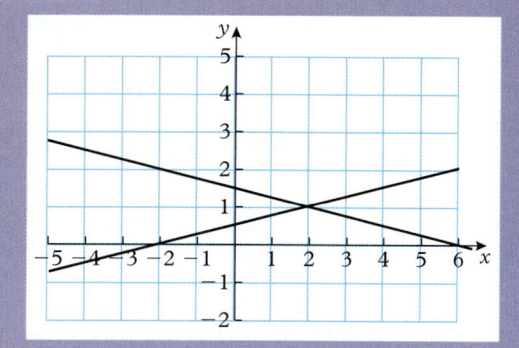

$x = $ ...... $y = $ ......   2 marks

Level 8

4 The diagram shows the graph with the equation $y = 2x^2$.

  a On the same axes draw the graph with the equation $y = x^2$.

  b On the same axes draw the graph with the equation $y = 2x^2 + 2$

  c On the same axes draw the graph with the equation $y = -2x^2$.

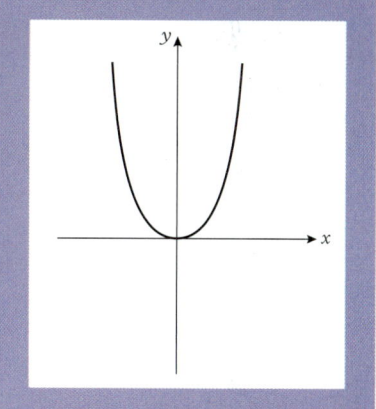

3 marks

5 The graph shows the journey of a car between two sets of traffic lights.

Describe in detail what happened during the journey of the car, as shown by the graph.

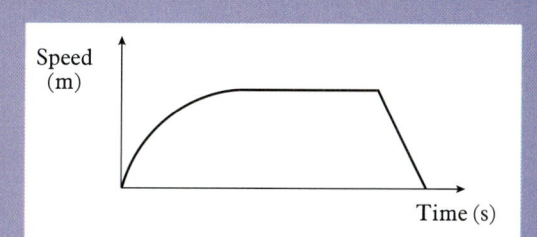

....

....

....

3 marks

71

WHAT YOU NEED TO KNOW

7 2D and 3D shapes

The triangle is used a lot in construction. It is a very strong shape.

Levels 3–5

- **Types of angle**

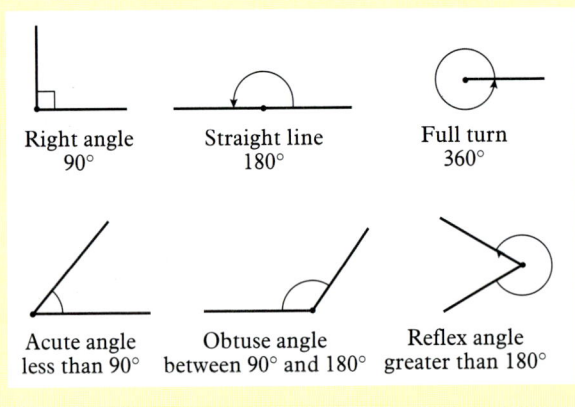

  Right angle 90°
  Straight line 180°
  Full turn 360°

  Acute angle less than 90°
  Obtuse angle between 90° and 180°
  Reflex angle greater than 180°

- **Points of the compass**

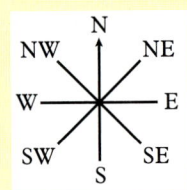

  N, NE, E, SE, S, SW, W, NW

- **Directions of turn**

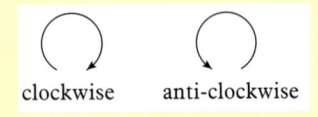

  clockwise   anti-clockwise

- **Angles on a straight line add up to 180°.**

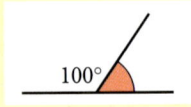

  100° the red angle must be 80° because 100 + 80 = 180

- **The angles in a triangle add up to 180°.**

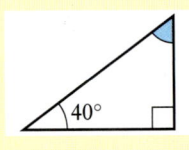

  40° the blue angle must be 50° because 90 + 40 + 50 = 180

WHAT YOU NEED TO KNOW

- **Types of triangle**

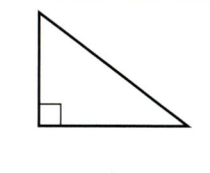

  Right-angled triangle

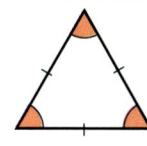

  Equilateral triangle
  All three sides are equal
  All three angles are equal

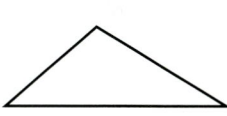

  Scalene triangle
  No equal sides
  No equal angles

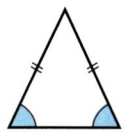

  Isosceles triangle
  Two sides are equal
  Two angles are equal

- **Constructing triangles**
  Always leave your construction lines on your diagram.

  Given three sides:
  Draw the longest side
  Use compasses to draw the two red arcs
  Join up the triangle

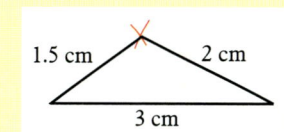

  Given one side and two angles:
  Draw the side
  Use a protractor to draw the two angles
  This will complete the triangle.

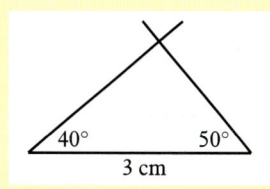

  Given two sides and one angle:
  Draw the longest side
  Use a protractor to draw the angle
  Mark the length of the other side
  Join up the triangle.

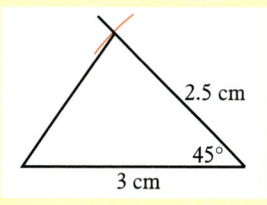

- **Two shapes are congruent if they are exactly the same.**
- **Parts of the circle**

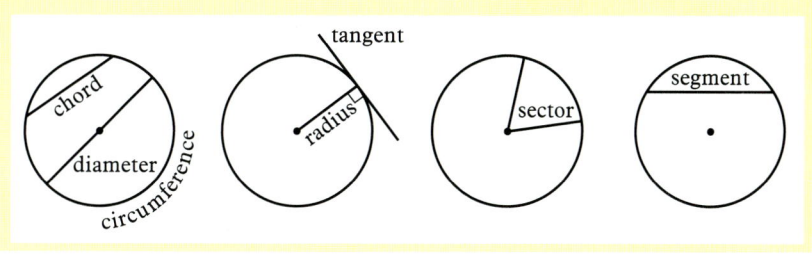

  chord, diameter, circumference, radius, tangent, sector, segment

WHAT YOU NEED TO KNOW

Level 6

- **Bearings are always measured clockwise starting from North.
  A bearing must always have three figures.**

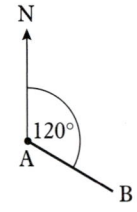

  Bearing of B from A = 120°

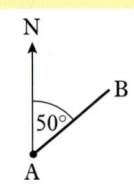

  Bearing of B from A = 050°

  Bearing of B from A = 210°

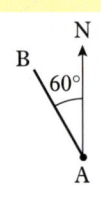

  Bearing of B from A = 300°

- **Constructions**

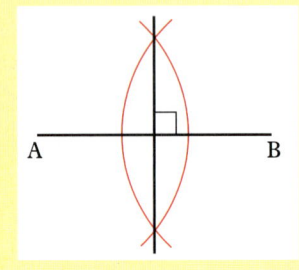

  Perpendicular bisector of AB

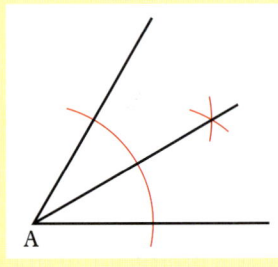

  Bisector of angle A

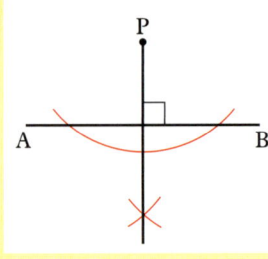

  Perpendicular from P onto AB

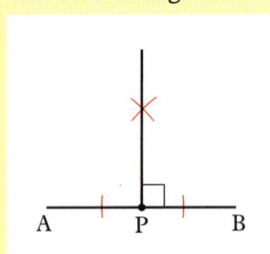

  Perpendicular at P

- **Constructing a right-angled triangle**
  Start by drawing a line.
  Construct an angle of 90° from a point on the line where you want the right angle to be.
  Use your compass to mark the base length with an arc.
  From this first arc use your compasses to draw a second arc to show where the hypotenuse meets the vertical.

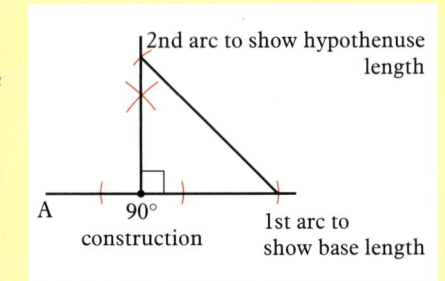

  2nd arc to show hypotenuse length

  90° construction

  1st arc to show base length

WHAT YOU NEED TO KNOW

- **Types of quadrilateral**
  A shape with four straight sides is a quadrilateral.

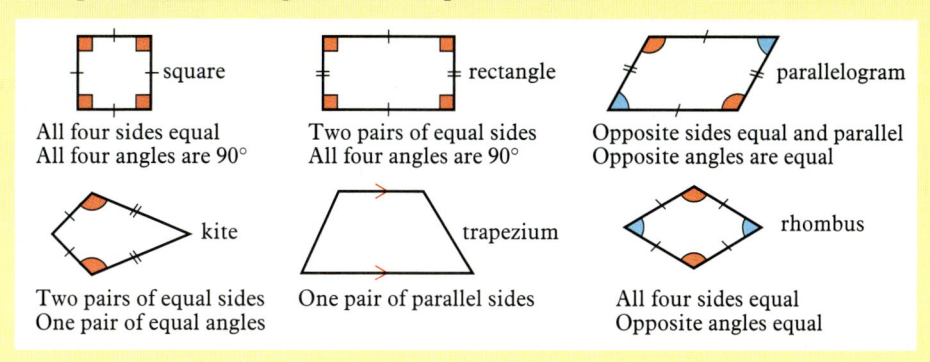

  square — All four sides equal. All four angles are 90°

  rectangle — Two pairs of equal sides. All four angles are 90°

  parallelogram — Opposite sides equal and parallel. Opposite angles are equal

  kite — Two pairs of equal sides. One pair of equal angles

  trapezium — One pair of parallel sides

  rhombus — All four sides equal. Opposite angles equal

- **Drawings of 3D shapes**

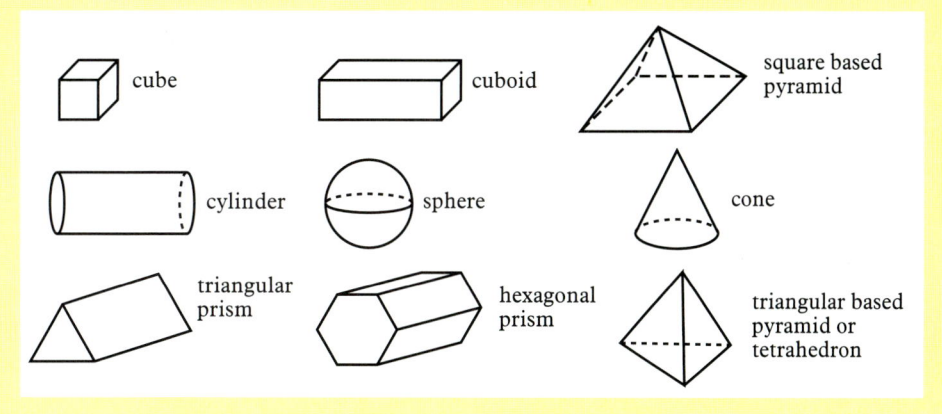

  cube, cuboid, square based pyramid, cylinder, sphere, cone, triangular prism, hexagonal prism, triangular based pyramid or tetrahedron

- When a solid is opened out and laid flat, the shape that you get is called a net of the solid.

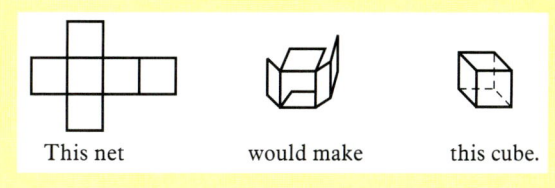

  This net   would make   this cube.

Plans and elevations

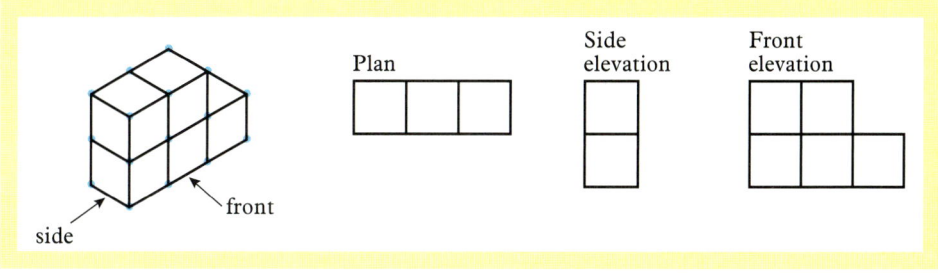

  side, front — Plan, Side elevation, Front elevation

75

WHAT YOU NEED TO KNOW

- **Names of polygons**

  3 sides – triangle   5 sides – pentagon   8 sides – octagon

  4 sides – quadrilateral   6 sides – hexagon   10 sides – decagon

- **A polygon is regular if all its sides are the same length and all its angles are equal.**

- **The angles of a quadrilateral add up to 360°.** 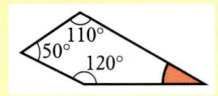

  The red angle is 80° because

  110 + 50 + 120 + 80 = 360

- **The exterior angles of a polygon add up to 360°.**

  If the hexagon is regular,

  each red angle = $\frac{360}{6}$ = 60° 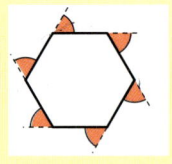

  The red angles add up to 360°

- **The angles inside a polygon are called interior angles.**

  They always make a straight line with the exterior angle.

  Exterior angle + Interior angle = 180°

  If the pentagon is regular, each red angle = $\frac{360}{5}$ = 72°

  so each blue angle = 180 − 72 = 108°

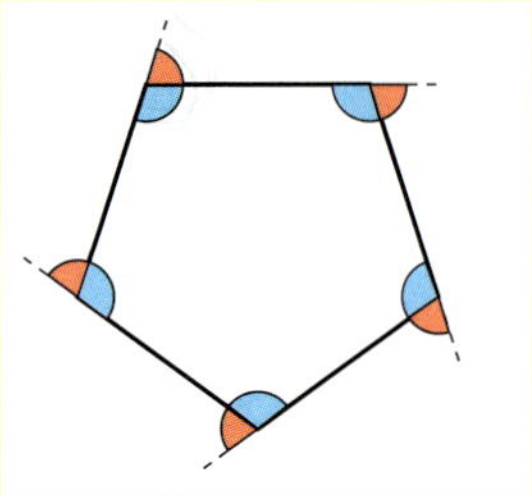

- **Angles between intersecting lines**

  The red angles are equal. The blue angles are equal. Each pair of angles is called vertically opposite angles or X angles.

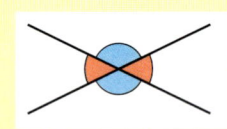

- **Angles between parallel lines**

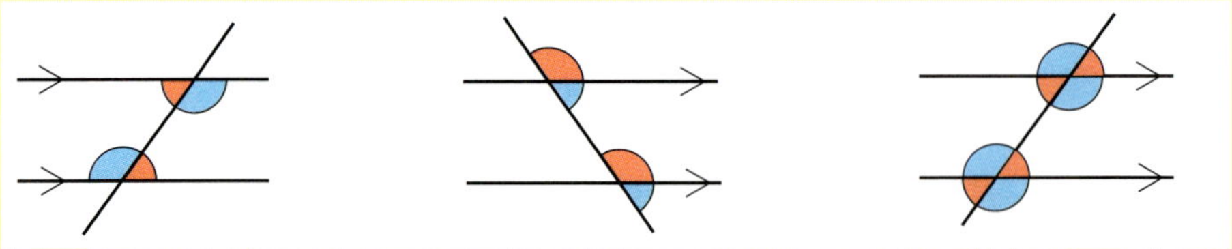

  The red angles are equal. The blue angles are equal. These angles are called **alternate angles** or Z angles.

  The red angles are equal. The blue angles are equal. These angles are called **corresponding angles** or F angles.

  All the red acute angles are equal. All the blue obtuse angles are equal. A red and a blue angle together add up to 180°.

TEST YOURSELF

Levels 3–5

1 Write down the name of each of these angles.

   a ....   **b** ....   **c** ....

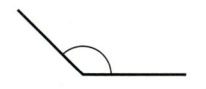

2 An aeroplane is flying due south. What direction will it be flying if it turns

   a 90° clockwise   ....   **c** 270° clockwise   ....

   b 45° anticlockwise   ....   **d** 135° anticlockwise   ....

3 Find the red angle.

   a  70° ....   **b** 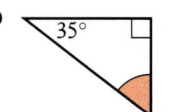 35° ....   **c** 80°, 20° ....

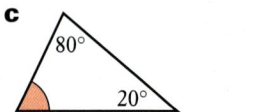

Level 6

4 Sketch each of these.
   a Cuboid   **b** Cylinder   **c** Square-based pyramid

5 Find
   a the red angle   **b** the blue angle

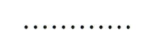

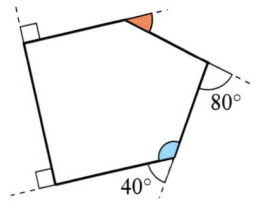

   ....   80°, 40°

6 Find each of the coloured angles.

   a red angle = ...

   b blue angle = ...

   c green angle = ...

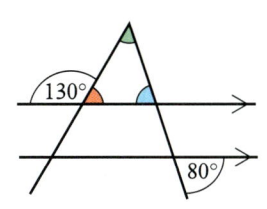

   130°, 80°

PRACTICE QUESTIONS

Levels 3–5

1 A cardboard box is being made for a small toy.
A plan of the **net** of the box is drawn first.

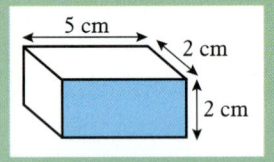

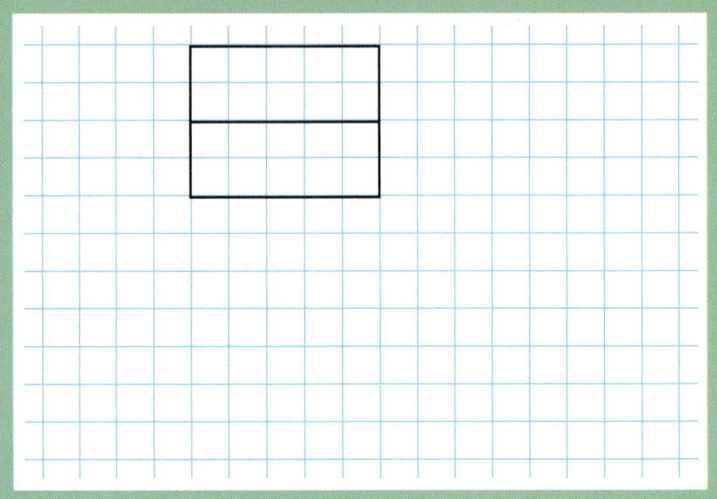

    a **Complete** the plan of the **net**.   3 marks

    b A **flap** is to be added so the box can have a **top** which closes.
       Add the **flap** to your plan of the **net**.   1 mark

2

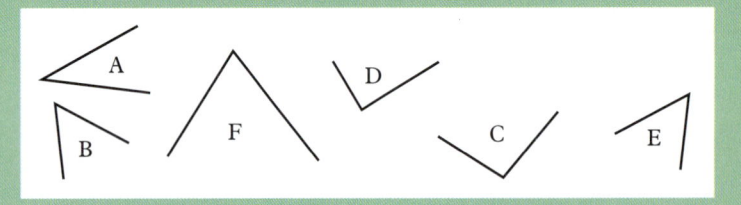

    a Which is the **smallest** angle?   .... 1 mark

    b Which is the **largest** angle?   .... 1 mark

    c Which **two** angles are the **same size**?   .... and .... 1 mark

    d Which angle is a **right angle**?   .... 1 mark

    e Draw an angle which is an **obtuse** angle.

                                                                                                1 mark

PRACTICE QUESTIONS

3 a John is facing **South**.
He turns **clockwise** through **2 right angles**.
In which direction will he now be facing?

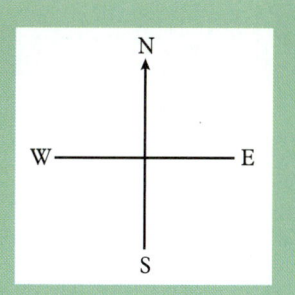

.... 1 mark

b Louise is facing **East**.
She turns **anticlockwise** through **3 right angles**.
In which direction will she now be facing?

.... 1 mark

Level 6

4 a

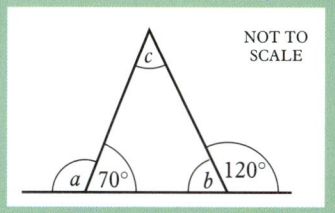

NOT TO SCALE

Calculate angles *a*, *b* and *c*.

a = ...   *b* = ...   *c* = ...   3 marks

b

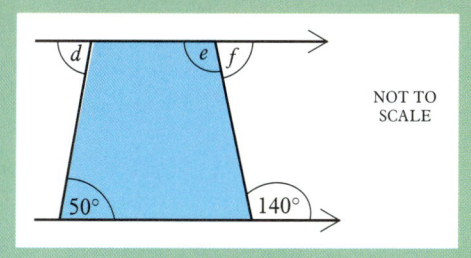

NOT TO SCALE

Calculate angles *d*, *e* and *f*.

d = ...   *e* = ...   *f* = ...   3 marks

What is the name of the blue shape?

.... 1 mark

PRACTICE QUESTIONS

5 This is an accurate scaled plan of a classroom.

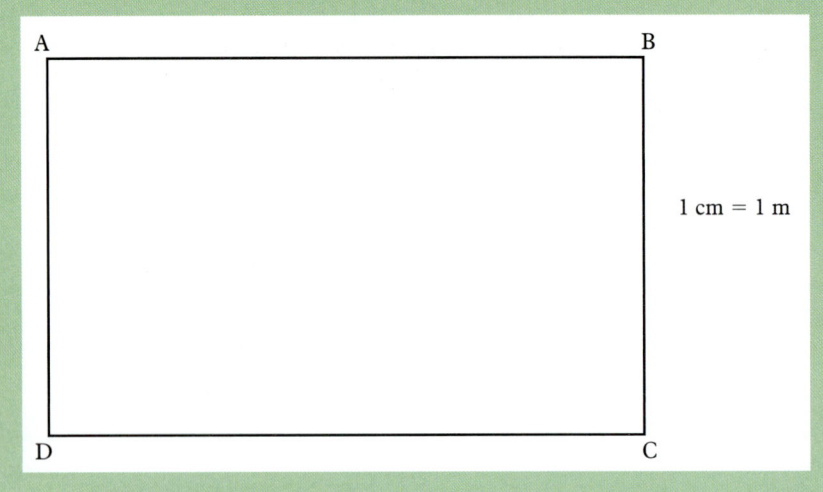

A B

1 cm = 1 m

D C

Using ruler and compasses only
- **a** construct a line to bisect angle D, and extend the line to meet AB
- **b** construct a line to bisect side DC, and extend the line to meet AB

Measure both lines, and write down the difference in their lengths.

.... 3 marks

6 In the space below, using ruler and compasses only, draw an accurate equilateral triangle with side of length 3.5 cm.

2 marks

WHAT YOU NEED TO KNOW

Level 7

- **Pythagoras' theorem is used to find the length of a side in a right-angled triangle.** You use it when you know the length of two of the sides.

- **The hypotenuse, h, is the longest side of a right-angled triangle.**
  Pythagoras' theorem tells you
  $$h^2 = a^2 + b^2$$

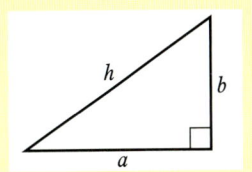

  The square of the hypotenuse = the sum of the squares of the other two sides

- **Finding the hypotenuse**
  $$h^2 = 8^2 + 11^2$$
  $$= 64 + 121$$
  $$= 185$$
  $$h = \sqrt{185}$$
  $$= 13.6 \text{ cm} \quad (3 \text{ sf})$$

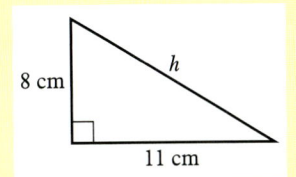

  Check that your answer for the hypotenuse is longer than each of the other two sides.

- **Finding one of the shorter sides**

  Always start by writing Pythagoras' theorem in the usual order starting with hypotenuse^2 = ...

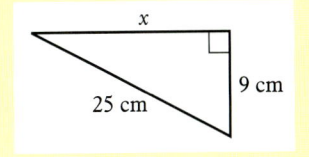

  $$25^2 = x^2 + 9^2$$
  $$25^2 - 9^2 = x^2$$
  $$x^2 = 625 - 81$$
  $$= 544$$
  $$x = \sqrt{544}$$
  $$= 23.3 \text{ cm} \quad (3 \text{ sf})$$

  Check that your answer is shorter than the hypotenuse.

WHAT YOU NEED TO KNOW

- **Finding the distance between two points**
  Draw a diagram showing the two points.
  Use the co-ordinates to find the lengths of the short sides of the triangle.
  Use Pythagoras to find the hypotenuse.
  $h^2 = 4^2 + 5^2$
  $ = 41$
  $h = 6.4$ units (1 dp)

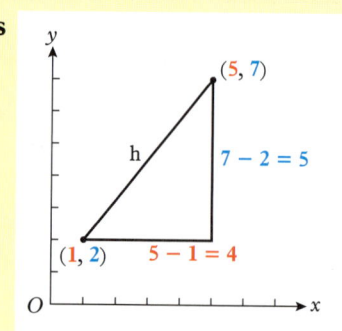

Level 8

- **Trigonometry is used to find the length of a side or an angle in a right-angled triangle.**

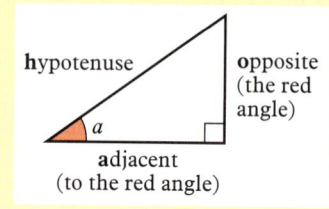

- **The three trig ratios are**
  $$\sin a = \frac{\text{opp}}{\text{hyp}} \quad \cos a = \frac{\text{adj}}{\text{hyp}} \quad \tan a = \frac{\text{opp}}{\text{adj}}$$

- **Finding a length**
  $\sin 40° = \dfrac{x}{5}$
  $5 \times \sin 40° = x$
  $ x = 3.21$ m (3 sf)

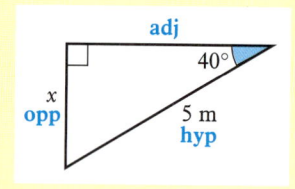

- **Finding an angle**
  $\tan a = \dfrac{12}{18}$
  $a = 33.7°$ (1 dp)

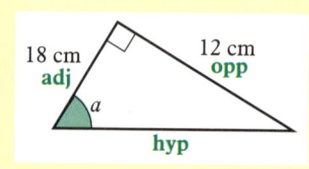

- **Bearings are often used in trigonometry questions.**
  Use the bearing to find the angle to work with.
  A ship travels 100 miles on a bearing of 204°. How far South does it travel?
  $\cos 24° = \dfrac{x}{100} \quad x = 91.4$ miles (3 sf)

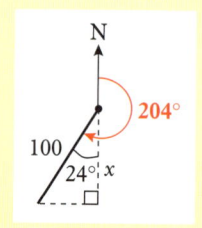

82

TEST YOURSELF

Level 7

1 Find x in each of these.

a
8 cm, 13 cm, x (right triangle with legs 8 and 13, hypotenuse x)

$x^2 = \ldots\ldots + \ldots\ldots$

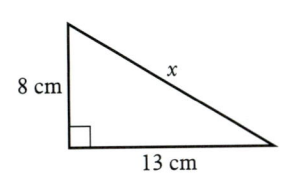

$ = \ldots\ldots$

$x = \sqrt{\ldots\ldots}$

$x = \ldots\ldots$ (3 sf)

b
7.8 cm, 12.3 cm, x (right triangle with one leg 7.8, hypotenuse 12.3, other leg x)

$\ldots\ldots = \ldots\ldots + \ldots\ldots$

$\ldots\ldots - \ldots\ldots = \ldots\ldots$

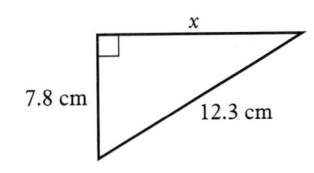

$x^2 = \ldots\ldots$

$x = \sqrt{\ldots\ldots}$

$x = \ldots\ldots$ (3 sf)

Level 8

2 Find p in each of these.

a 14.1 m, 38°, p

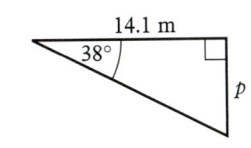

$\ldots\ldots 38° = \dfrac{p}{\ldots\ldots}$

$\ldots\ldots\ldots = p$

$p = \ldots\ldots$ m (3 sf)

b 17 cm, 11 cm, p

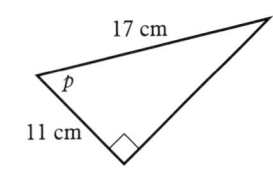

$\ldots\ldots p = \dfrac{\ldots\ldots}{\ldots\ldots}$

$p = \ldots\ldots$ ° (3 sf)

3 A plane flies 500 miles on a bearing of 306° from A. How far West does it travel?

....

....

....

PRACTICE QUESTIONS

Levels 5–7

1 This shape is a **square-based** pyramid.

Write down the letter of the **one** net which can be used to make this pyramid.

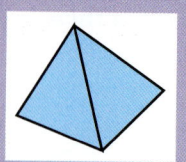

A B C D E F

1 mark

2 A boat is to sail around a hexagonal course ABCDEF, past six buoys. The boat starts from point A.

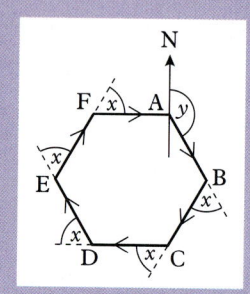

a Through what angle *x* should the boat turn at each of the buoys?

$x =$ .... 1 mark

b What bearing *y* should the boat sail on at the start of the race?

$y =$ .... 1 mark

3 Calculate the area of the triangle.

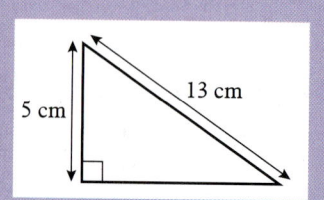

Area = ... cm²   3 marks

PRACTICE QUESTIONS

Level 8

5 A boat is sailed around a lake.

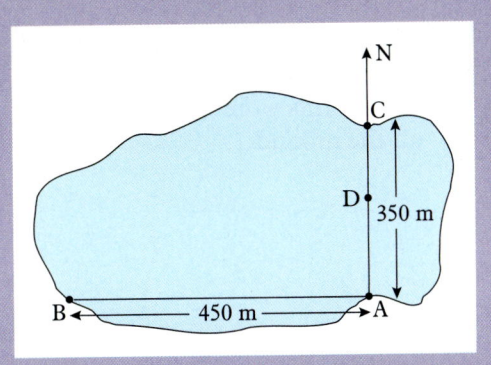

Point C is 350 metres due North of A.
Point B is 450 metres due West of A.

a Calculate the direct distance from point B to point C.

.... m   2 marks

b Calculate the bearing the boat must sail on to travel from point B to point C.

.... °   2 marks

c Calculate the bearing the boat must sail on to travel from point C to point B.

.... °   1 mark

d The boat sails on a bearing of 60° from point B, across the lake until it reaches point D, due North of A.
Calculate the distance from B to D.

.... m   2 marks

WHAT YOU NEED TO KNOW

8 Position and movement

The window acts like a mirror. The reflection makes it look as though this woman has both feet off the ground.

Levels 3–5

- **A line of symmetry divides a shape into two identical halves. Each half is a reflection of the other.**
  If you fold a shape along a line of symmetry each half fits exactly on top of the other.

- **A shape can have more than one line of symmetry or none at all.**

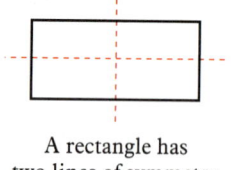

   A rectangle has two lines of symmetry

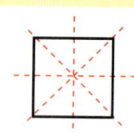

   A square has four lines of symmetry

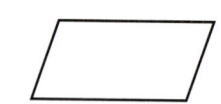

   A parallelogram does not have any lines of symmetry

   The dashed red lines are the lines of symmetry.
   They are sometimes called mirror lines.

- **To reflect a shape in a mirror line**
   (1) Reflect each corner in turn. The new position must be the same distance on the other side of the mirror line. Use the squares to help you.
   (2) Join these points to get the reflected shape.

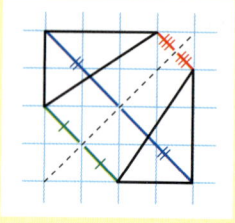

WHAT YOU NEED TO KNOW

- **You can reflect 3D shapes in a mirror.**
  This shape is symmetrical about the mirror.

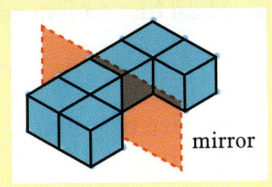

- **A shape has rotational symmetry if it fits on top of itself more than once as it makes a complete turn.**
- **The order of rotational symmetry is the number of times the shape fits on top of itself.**
- **The centre of rotation is the point at the centre of the shape that stays still as the shape turns.**

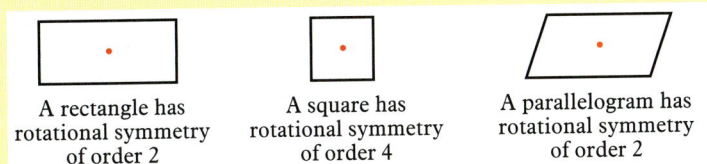

  A rectangle has rotational symmetry of order 2

  A square has rotational symmetry of order 4

  A parallelogram has rotational symmetry of order 2

- **To rotate a shape about any point.**
  (1) You need to know the angle and the direction of turn.
  (2) Trace the shape.
  (3) Put your pencil on the centre of rotation and rotate the tracing paper.
  (4) Draw the shape in its new position.

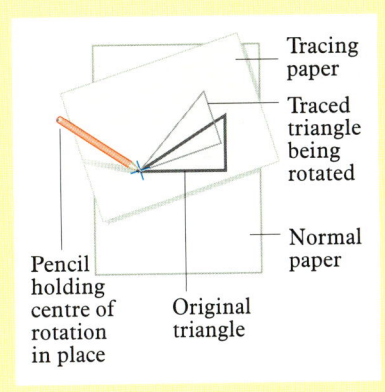

  You can use a cross on the tracing paper to help you see when you have rotated through 90°, 180°, 270°.

- **A translation is a movement in a straight line.**
  The shape A has been moved three squares to the right and one square down to get to B.
  You can write this as $\begin{pmatrix} 3 \\ -1 \end{pmatrix}$

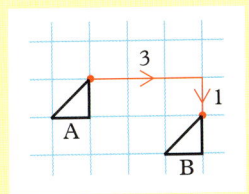

WHAT YOU NEED TO KNOW

- **When you are doing more than one transformation on a shape make sure you do them in the right order.**

- **An enlargement changes the size of a shape.**
  The change is the same in all directions.

- **The scale factor tells you how many times bigger the enlargement is.**
  Rectangle B is an enlargement of rectangle A, scale factor 2.

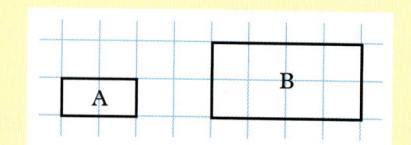

- This map shows some streets.
  The real streets are an enlargement of the map.
  The scale of the map tells you the scale factor of the enlargement.
  The distance between the church and the school on the map is 2 cm.
  To find the actual distance multiply by the scale factor

  Actual distance = 2 × 10 000 cm
        = 20 000 cm
        = 200 m

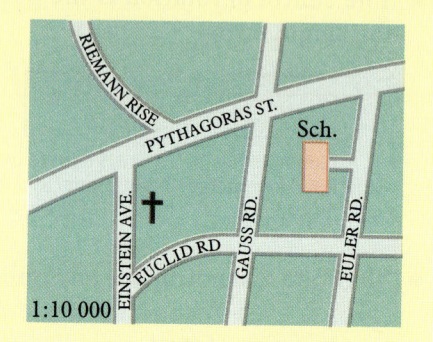

Level 6

- **An enlargement can be done from a point.**
  This point is called the centre of enlargement, C.
  Look at the corner of the small triangle marked ●
  ● is 1 square from C.
  The new position will be 1 × 3 = 3 squares from C.
  Do the same for the other corners of the shape.

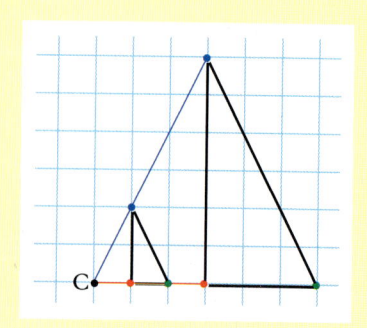

WHAT YOU NEED TO KNOW

- **Symmetry of the regular polygons.**

  Equilateral triangle

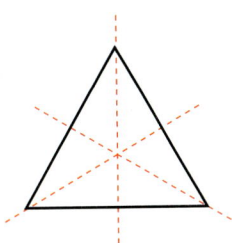

  three lines of symmetry
  rotational symmetry
  of order 3

  Square

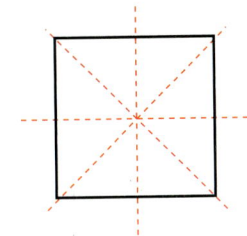

  four lines of symmetry
  rotational symmetry
  of order 4

  Regular pentagon

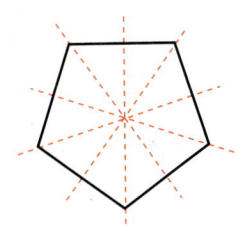

  five lines of symmetry
  rotational symmetry
  of order 5

  Regular hexagon

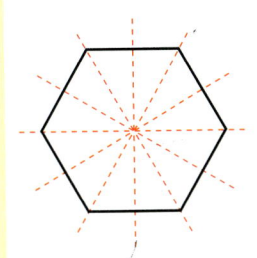

  six lines of symmetry
  rotational symmetry
  of order 6

  Regular octagon

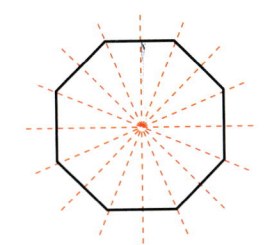

  eight lines of symmetry
  rotational symmetry
  of order 8

  Regular decagon

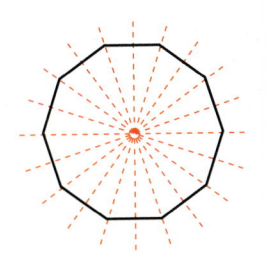

  ten lines of symmetry
  rotational symmetry
  of order 10

- **Symmetry of solids**
  This is a cuboid.
  You can see the three possible places to put a mirror so that the two halves are symmetrical.
  When you put a mirror into a shape like this the mirror is called a plane of symmetry.

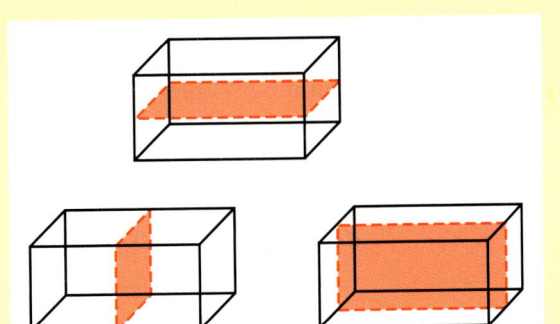

TEST YOURSELF

Levels 3–5

1 For each of these shapes draw the lines of symmetry and write down the order of rotational symmetry.

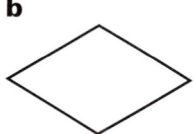

   a   b   c   d

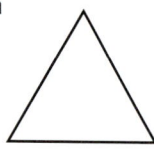

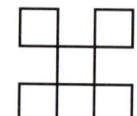

   order ......   order ......   order ......   order ......

2 **a** Translate triangle A 6 squares to the right and 2 squares up.
   b Rotate triangle A 90° clockwise about •
   c Reflect triangle A in the blue line.

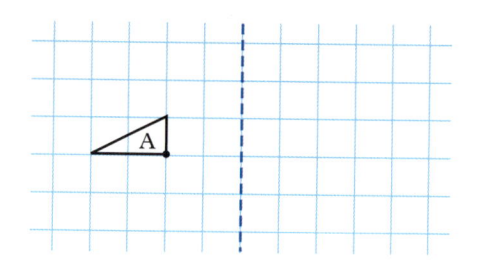

Level 6

3 Enlarge this shape using centre C and scale factor 2.

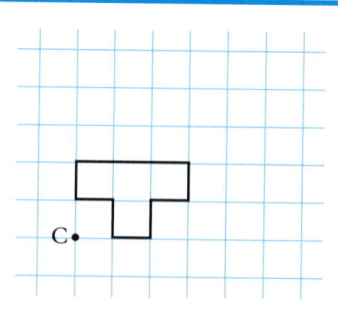

4 Write down the number of planes of symmetry.

   a   b   c

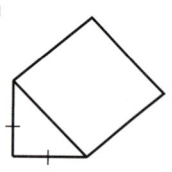

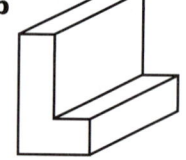

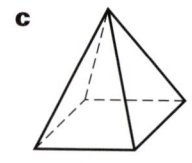

                                             square-based pyramid

   ....

90

PRACTICE QUESTIONS

Levels 3–5

1 a Shade in **one more square** to make a shape which has line AB as a **line of symmetry**.

1 mark

b Shade in **one more square** to make a shape with line CD as a **line of symmetry**.

1 mark

c Shade in **three more squares** to make a shape which has **two lines of symmetry**.

1 mark

d Shade in **one more square** to make a shape which has **rotational symmetry**.

1 mark

e Shade in three more squares to make a shape which has **two lines of symmetry** and **rotational symmetry** about the marked point.

1 mark

f Shade A has been **rotated clockwise** onto shape B. Mark with a clear cross the **centre of rotation**.

Write down the angle of rotation.

....

2 marks

2 For each of these patterns draw in **all** the **lines of symmetry**.

a   b   c

3 marks

PRACTICE QUESTIONS

3 Reflect this triangle in the mirror line.

2 marks

Level 6

4 Write down the number of **planes of symmetry** in the following solids.

a Cube ………

b Prism ………

2 marks

5 Draw an **enlargement** of this shape by a **scale factor** of 2.

3 marks

6 Draw an **enlargement** of this shape by a **scale factor** of 3 from the centre of enlargement A.

•A

2 marks

WHAT YOU NEED TO KNOW

Level 7

- **An enlargement can also make a shape smaller.**
  This happens when the scale factor is between 0 and 1. Shape A has been enlarged by scale factor $\frac{1}{2}$.

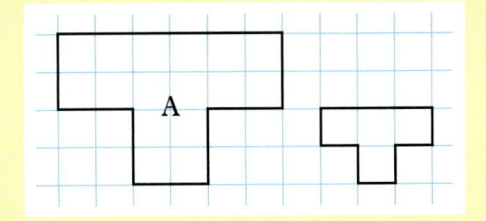

- **The locus of an object is all the possible positions that the object can take as it moves according to a rule.**
  You can describe a locus in words or with a diagram.
  The red line shows the locus in each of these.

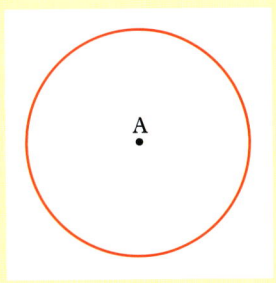

Points that are 1.5 cm from A.

A circle, centre A radius 1.5 cm.

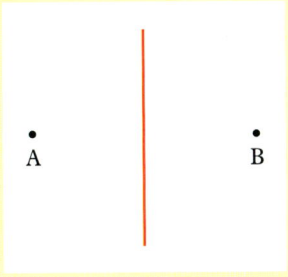

Points that are equidistant from the two points A and B.

The perpendicular bisector of the line AB.

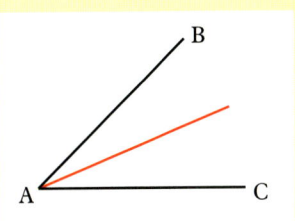

Points that are equidistant from the lines AB and AC.

The bisector of angle BAC.

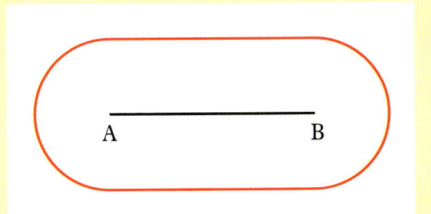

Points that are 1 cm from the line AB.

Pair of parallel lines with a semicircle at each end.

93

WHAT YOU NEED TO KNOW

- **You can also shade regions to show where something can be.**
  The red region shows the points that are nearer to A than to B.

Level 8

- **Two triangles are similar if they have the same angles.**
  Triangles P and Q are similar.

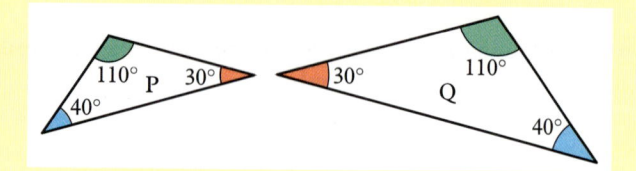

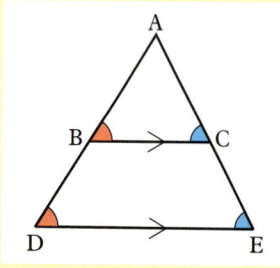

The red angles are corresponding angles so they are equal.
The blue angles are corresponding angles so they are equal.
Triangles ABC and ADE are similar.

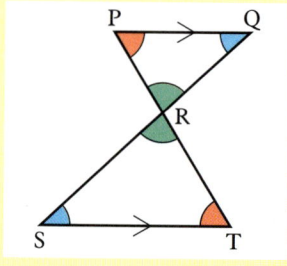

The red and blue angles are alternate angles so they are equal.
The green angles are opposite angles and so they are equal.
Triangles PQR and TSR are similar.

- **Similar triangles are enlargements of each other.**
  Use the scale factor to find missing lengths.
  Make sure you're working with corresponding sides.

  Triangles CDE and FGE are similar.
  GE and ED are corresponding sides as they are both opposite the red angles.
  The scale factor for the enlargement from FGE to CDE is 6 ÷ 4 = 1.5.
  GF and CD are corresponding sides
  Length CD = 3 × 1.5
  = 4.5 cm.

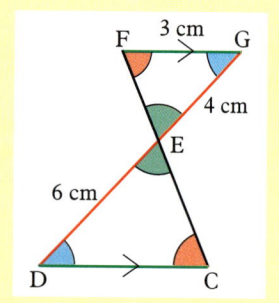

WHAT YOU NEED TO KNOW

- **Congruent**
  Two shapes are congruent if they are identical.
  Use one of these rules to show that two triangles are congruent.

  Rule 1 All three pairs of sides are equal

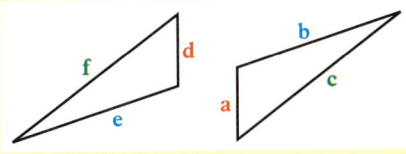

  Here $a = d$
  $b = e$
  and $c = f$

  Remember this rule as **SSS**.

  Rule 2 Right angle, hypotenuse and side are equal

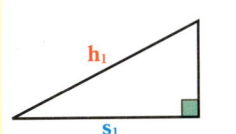

  Here $h_1 = h_2$
  $s_1 = s_2$
  and both triangles have a right angle

  Remember this rule as **RHS**.

  Rule 3 Two pairs of corresponding sides are equal and the angles *between* the pairs of sides are also equal

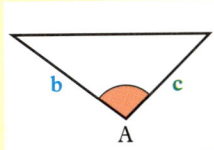

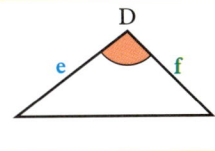

  Here $\hat{A} = \hat{D}$
  $b = e$
  and $c = f$

  Remember this rule as **SAS**.
  You write the A in the middle to remind you that the equal angles must be *between* the two sides.

  Rule 4 Two pairs of angles are equal and a pair of corresponding sides is also equal

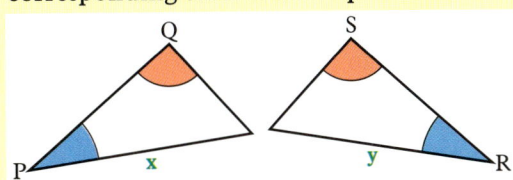

  Here $\hat{Q} = \hat{S}$
  $\hat{P} = \hat{R}$
  and $x = y$

  Remember this rule as **AAS**.
  You must remember that the equal sides must be in the same place.

Once you have proved that two triangles are congruent remember that all corresponding sides and angles are equal.

TEST YOURSELF

Level 7

1 Enlarge the shape using centre C and scale factor $\frac{1}{2}$.

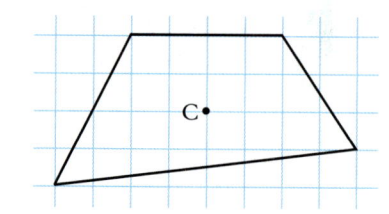

2 The diagram shows a wheel.
   a Draw the locus of the red point on the rim of the wheel as the wheel rolls along the line.
   b Draw the locus of the blue point at the centre of the wheel.

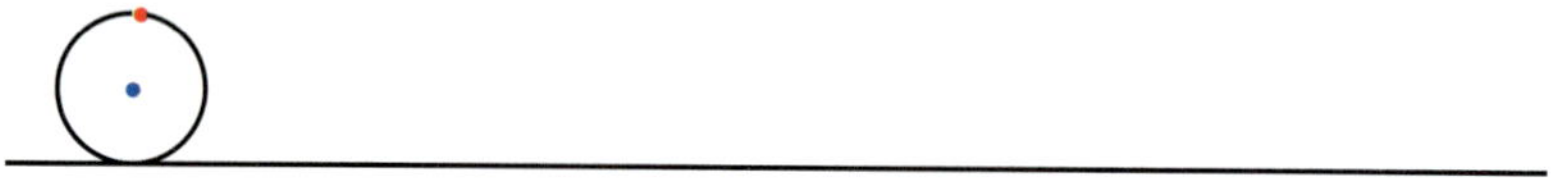

3 An ambulance station must be less than 4 km from the hospital and less than 3 km from the motorway. Show the region where it could be.

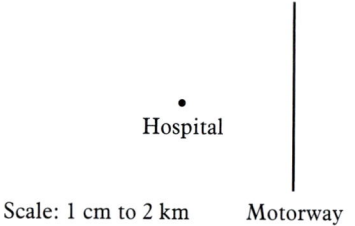

Scale: 1 cm to 2 km

Hospital

Motorway

Level 8

4 a Mark the pairs of equal angles.
   b Write down the corresponding side to

   (1) PQ ......   (2) SR ....

   c Find the scale factor for the enlargement from RST to PQR.

   ...... ÷ ...... = ...

   d Find the length SR.

   ...... ÷ ...... = ... cm

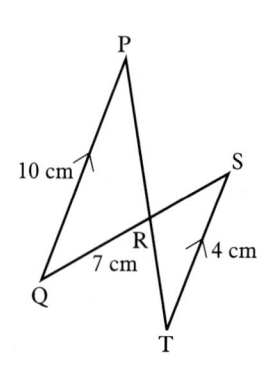

10 cm, 7 cm, 4 cm
P, Q, R, S, T

PRACTICE QUESTIONS

Levels 5–7

1 This is a logo measuring 7 cm by 10 cm.
June wants to use a photocopier to reproduce the logo on three different sizes of paper. The logo needs to be as large as possible on each sheet of paper. Find the scale factor she should use for each size of paper.
Give your answers correct to 2 dp.

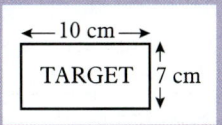

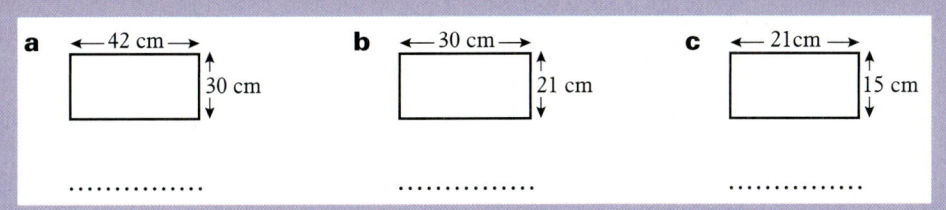

TARGET — 10 cm × 7 cm

a 42 cm × 30 cm

b 30 cm × 21 cm

c 21 cm × 15 cm

....

3 marks

2

A — 6 m — B, 3 m, D C

This is a plan of a garden plot.
Draw an **accurate scale plan** of the plot.
Use a scale of 1 square to represent 1 m. *1 mark*

a A row of flowers is planted in a **straight line** exactly **1.5 m** from the side BC. Show the location of the line of flowers on your scale plan. *1 mark*

b The plot is divided by a small fence. The fence starts in corner D, and runs across the plot so as to be the same distance from side AD and side DC. On your scale plan show the position of the fence. *1 mark*

c Seeds are sown in such a way as to be no more than 1 metre from corner A. On your scale plan shade the location in which the seeds can be sown. *1 mark*

PRACTICE QUESTIONS

3 This is an accurate plan of a garden pond.
It has been drawn using a scale of 1 cm to represent 1 m.

  a A fence is to be put up which is **exactly 1 metre** from the edge of the pond. Draw **accurately** the position of the fence on the scale plan. *3 marks*

  b A small fountain is at the corner A of the pond. It sprays water up to one metre away. On the scale plan shade the part of the pond which is sprayed by the fountain. *1 mark*

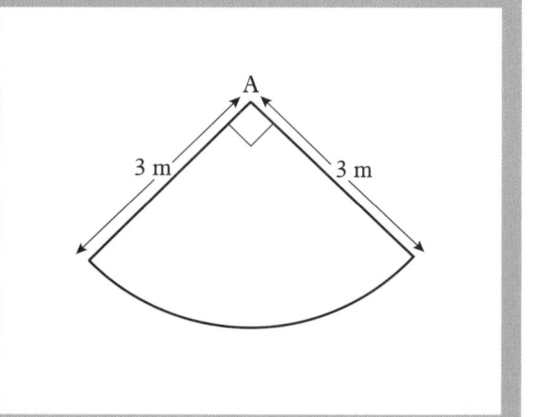

Level 8

4 These two tins are mathematically similar. The external dimensions are shown. Calculate the height h of the larger tin.

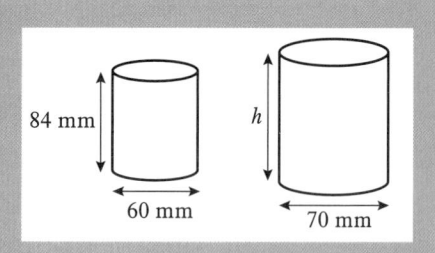

$h = $ .... mm *2 marks*

5 A 10 metre high pole is held in place by two metal struts, AE and BC.
The position of the struts is described in the diagram.
AD = 10 m, BC = 6 m, DE = 12 m,
AC = 8.5 m
BC is parallel to the level ground DE.
Calculate the length of CE.

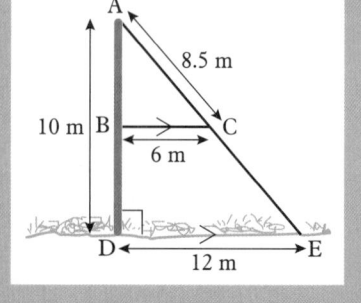

CE = ... m *2 marks*

PRACTICE QUESTIONS

6

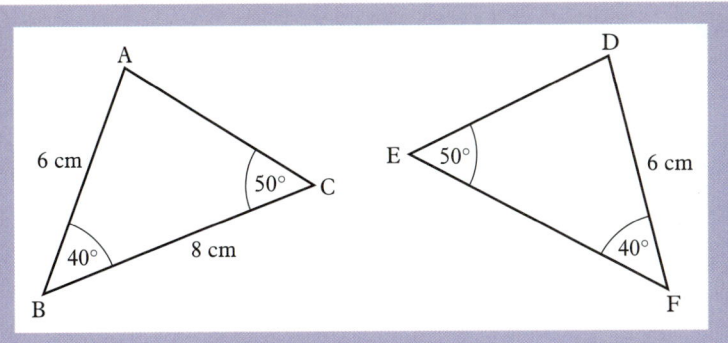

ABC and DEF are two congruent triangles.

   a Explain why ABC and DEF are congruent.   1 mark

   b Use both triangles to find the length DE.   3 marks

7

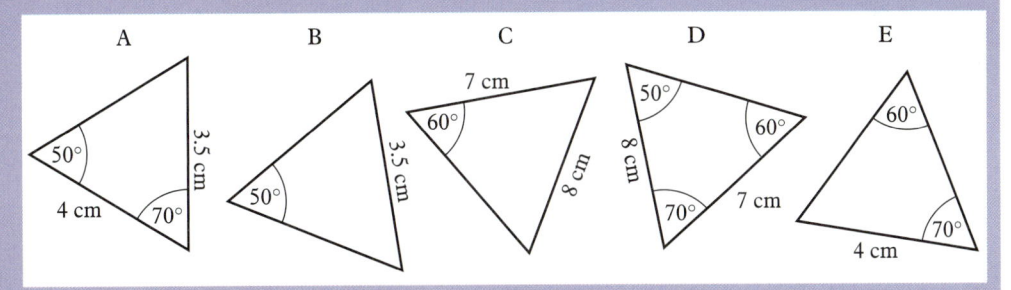

   a Write down the letters of two triangles that are mathematically **similar** to each other, but **not** congruent.

      ....

      Explain why they are only similar.   2 marks

   b Write down the letters of two triangles that are **congruent** to each other.

      ....

      Explain why they are congruent.   2 marks

   c Write down the letters of two **different** triangles that are **congruent** to each other.

      ....

      Explain why they are congruent.   2 marks

WHAT YOU NEED TO KNOW

9 Units of measurement

French scientists worked out the distance from the North Pole to the Equator and divided by 10 million. They called this distance 1 metre.

Levels 3–5

- **The metric units of length are millimetres (mm), centimetres (cm), metres (m) and kilometres (km).**

  10 mm = 1 cm   100 cm = 1 m   1000 m = 1 km

- **The metric units of capacity are millilitres (ml), centilitres (cl) and litres (l).**

  10 ml = 1 cl 100 cl = 1 l 1000 ml = 1 l

- **The metric units of mass are milligrams (mg), grams (g), kilograms (kg) and tonnes (t).**

  1000 mg = 1 g   1000 g = 1 kg   1000 kg = 1 t

- **To convert between units you need to know the above conversion facts.**

  Convert 4800 g to kilograms.
  Look for the conversion fact that you need.
  To go from the larger unit to the smaller unit you multiply.
  To go from the smaller unit to the bigger unit you divide.

  1000 g = 1 kg
  g → kg
  smaller → bigger
  so divide by 1000

  4800 g ÷ 1000 = 4.8 kg

WHAT YOU NEED TO KNOW

- **The Imperial units of length are inches (in), feet (ft), yards (yd) and miles.**

  12 in = 1 ft   3 ft = 1 yd   1760 yd = 1 mile

- **The Imperial units of mass are ounces (oz), pounds (lb) and stones (st).**

  16 oz = 1 lb   14 lb = 1 st

- **The Imperial units of capacity are fluid ounces (fl oz), pints (pt) and gallons (gal).**

  20 fl oz = 1 pt   8 pt = 1 gal

- **Converting between metric and Imperial units of length**

  You need to remember these conversion numbers and when to multiply and divide.

	Conversion number
1 in is about 2.5 cm	2.5
1 ft is about 30 cm	30
1 yd is about 90 cm	90
1 mile is about 1.6 km	1.6

  Imperial ⇄ metric (multiply / divide)

  Convert 3 yd to cm
  Look for the conversion number that you need.

  3 yds = 3 × 90 = 270 cm

  yd → cm
  90
  Imperial → metric
  so multiply

- **Converting mass and capacity**

	Conversion number
1 oz is about 30 g	30
1 lb is about 450 g	450
1 st is about 6.5 kg	6.5
1 pt is about 600 ml	600
1 gal is about 4.5 l	4.5

- **Converting from metric to Imperial units**

  You may find it easier to remember these:
  1 metre is about 3 feet.   8 kilometres is about 5 miles.
  1 kilogram is about 2.2 pounds.   1 litre is just less than 2 pints.

101

WHAT YOU NEED TO KNOW

- **To read a scale**

  (1) Work out what each division represents.
  (2) Work out the reading.

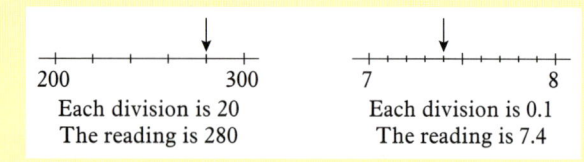

  Each division is 20
  The reading is 280

  Each division is 0.1
  The reading is 7.4

- **Time can be written using am and pm or the 24-hour clock**

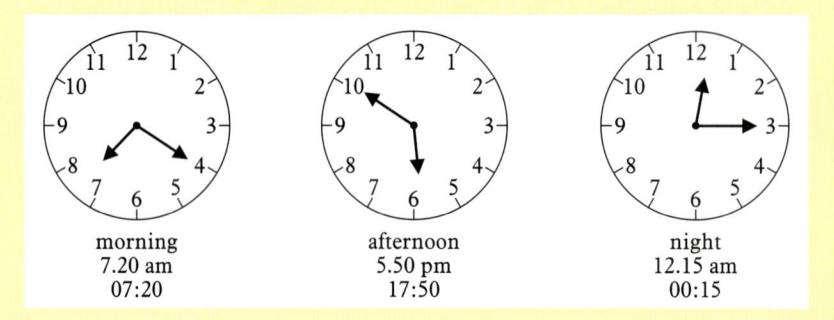

  morning
  7.20 am
  07:20

  afternoon
  5.50 pm
  17:50

  night
  12.15 am
  00:15

- You need to be able to estimate in both Imperial and metric units. Remember these familiar objects to help you.

  Length
  1 in
  1 cm
  2 m
  6 ft
  1 ft
  30 cm

  Mass
  30 g or 1 ounce
  1 kg or 2.2 lb
  100 kg or 15 st

  Capacity
  250 ml or ½ pt
  2 l or 3½ pt
  90 l or 20 gal

TEST YOURSELF

Levels 3–5

1
- **a** 1 m = …… cm
- **b** …… in = 1 ft
- **c** …… oz = 1 lb
- **d** 1 kg = …… g
- **e** …… pt = 1 gal
- **f** 1 l = …… ml
- **g** 2 cm = …… mm
- **h** …… kg = 3 t
- **i** 400 cl = …… l
- **j** …… oz = 3 lb
- **k** 6 yd = …… ft
- **l** 24 in = …… ft

2
- **a** 1 mile is about …… km
- **b** 1 pt is about …… ml
- **c** 1 yd is about …… cm
- **d** 1 gal is about …… l
- **e** 1 oz is about …… g
- **f** 1 lb is about …… g

3
- **a** Each division is …
  The reading is …
- **b** Each division is …
  The reading is …
- **c** Each division is …
  The reading is …

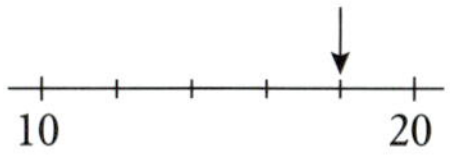

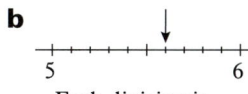

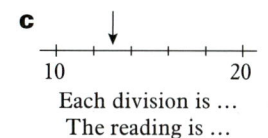

4
- **a** 6:50 am = ……
- **b** 23:10 = ……
- **c** 01:45 = ……
- **d** 4:36 pm = ……

5
- **a** length is about …… m
- **b** mass is about …… oz
- **c** width is about …… in
- **d** capacity is about …… ml

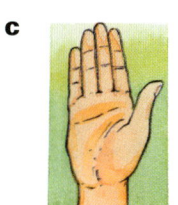

PRACTICE QUESTIONS

Levels 3–5

1   BUS TIMETABLE

Underhill	depart	07:30	08:15	09:00	09:45	10:30
Bramley	arrive	08:05	08:50	09:35	10:20	11:05

a A bus leaves **Underhill** at **09:45**.
   At what time does it arrive at **Bramley**?   ....   1 mark

b A bus arrives at **Bramley** at **10:20**.
   At what time did it leave **Underhill**?   ....   1 mark

c How long does the bus journey take from **Underhill** to **Bramley**?

                                                               .... minutes   1 mark

d Which buses could you catch to get to **Bramley** by **10:00**?

     .... or .... or ....   1 mark

e Vincent needs to be in **Bramley** by **09:00**.
   What is the time of the **latest** bus he can catch from **Underhill**?

                                                               ....   1 mark

f The last evening bus from Underhill is at **23:15**.
   Draw hands on this clock face to represent this time.

                                                                    1 mark

2

a Write down the **length** of the pencil

     in centimetres ....   in millimetres ....   2 marks

b Another pencil of length 5.3 cm has 16 mm cut from its length.
   What is the new length of the pencil?
   Give your answer in centimetres.   .... cm   1 mark

PRACTICE QUESTIONS

3 a A piece of cheese is weighed.
Write down the mass shown
by the scale.

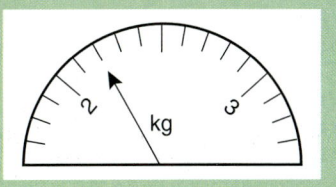

.... kg   1 mark

b 600 g of cheese is cut off.
How much cheese remains?
Give your answer in kilograms.   .... kg   2 marks

4 Write down your estimate of each of the following:

a The mass of a bag of sugar   .... kg

b The capacity of a milk bottle   l

c The length of an ant   .... mm

5 The conversion graph shown on the grid is for centimetres and inches.

a Change **2 inches** into **centimetres**.

.... cm

b Change **20 centimetres** into **inches**.

.... inches

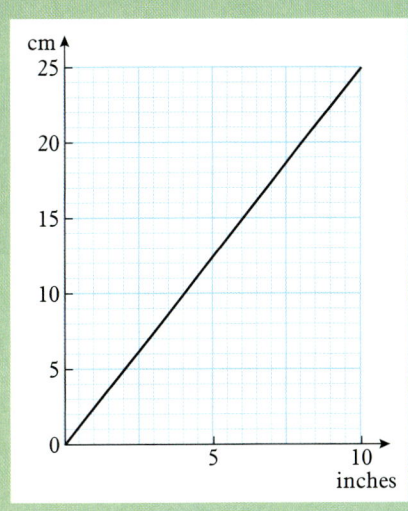

2 marks

105

WHAT YOU NEED TO KNOW

Level 7

- **Lower and upper limits**
  Any number in the red part of the number line rounds to 7 to the nearest whole number.

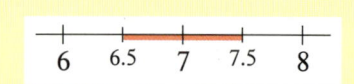

  6.5 is called the lower limit.
  7.5 is called the upper limit.

- **S**peed = $\dfrac{\textbf{D}\text{istance}}{\textbf{T}\text{ime}}$ **T**ime = $\dfrac{\textbf{D}\text{istance}}{\textbf{S}\text{peed}}$ **D**istance = **S**peed × **T**ime

  This triangle can help you to remember these.
  Cover up the letter you're trying to find.
  What is left is the rest of the equation.

  An object moves 300 m in 2 mins
  Find the average speed in metres per second.
  To get the speed in metres per second you must work in metres and seconds.
  So the object moves 300 m in 120 s.

  Speed = $\dfrac{300}{120}$ = 2.5 m/s.

- **Travel graphs**
  The red part shows movement away from home. 50 km in 2 hours is a speed of 25 km/h.
  The blue part shows a stop lasting 1 hour.
  The green part shows the journey home.
  There is a further 1 hour stop 30 km from home.

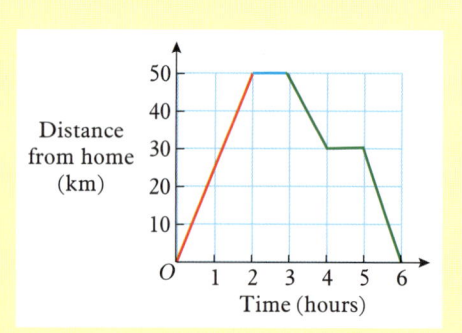

  The average speed = total distance ÷ total time = 100 ÷ 6 = $16\frac{2}{3}$ km/h

- **D**ensity = $\dfrac{\textbf{M}\text{ass}}{\textbf{V}\text{olume}}$ **V**olume = $\dfrac{\textbf{M}\text{ass}}{\textbf{D}\text{ensity}}$

  Mass = **D**ensity × **V**olume.

106

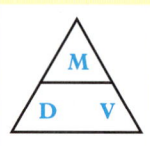

TEST YOURSELF

Level 7

1 The sides of this box are correct to the nearest whole number.

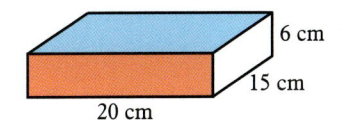

   a The lower limits of each length are

     …… cm, …… cm, …… cm.

   b The upper limits of each length are …… cm, …… cm, …… cm.

2 The travel graph shows Tim's cycle journey from Cranley to Marsh.

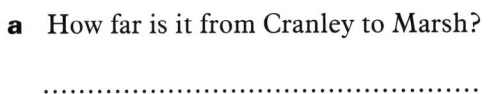

   a How far is it from Cranley to Marsh?

   b How long did Tim stop for a rest?

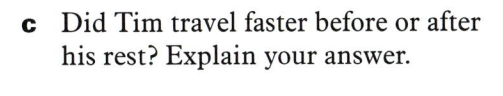

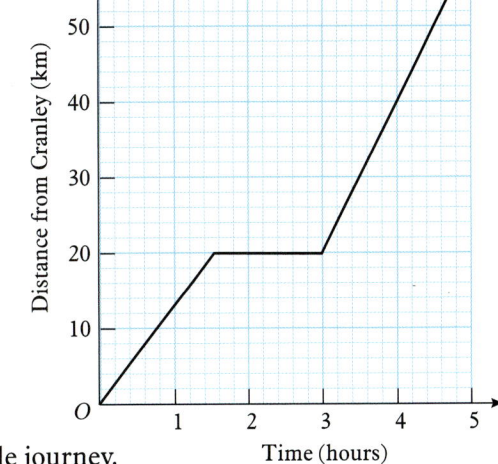

   c Did Tim travel faster before or after his rest? Explain your answer.

   d Find Tim's average speed for the whole journey.

3 Complete the table.

	Speed	Distance	Time
a		13 m	2 s
b	5 m/s		2.5 mins
c	12 km/h	4 km	
d	mph	40 miles	$1\frac{1}{4}$ hours

4 Complete the table.

	Density	Mass	Volume
a		50 g	10 cm^3
b	50 kg/m^3	100 kg	
c	15 g/cm^3		20 cm^3
d	g/cm^3	2.5 kg	100 cm^3

PRACTICE QUESTIONS

Levels 5–7

1 Estimate the readings on the following scales.
Give the units of measurement with your answer.

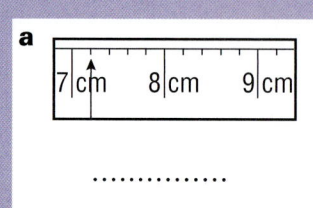

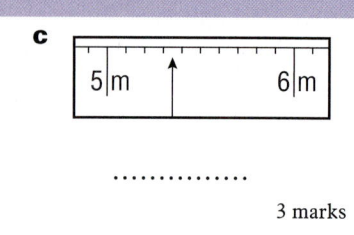

a

b

c

....

....

....

3 marks

2 The scales show the mass of a man in stones.

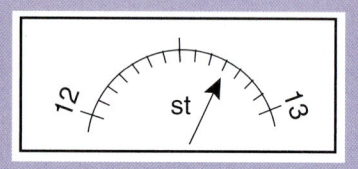

a Write down, in **stones and pounds**, the reading on the scales.

.... st .... lb   1 mark

b The man has lost $1\frac{1}{2}$ stones while on a diet.
How much did the man weigh **before** he went on the diet?

.... st .... lb   1 mark

3 A 600 l tank is being filled with water.
The table shows some readings taken while the tank is being filled.

Time (min)	1	6	10	15	25	30	
Capacity (litres)		72		180			600

a Complete the table. 2 marks

b C is the capacity of the tank in litres,
and T is the time in minutes.
Write an equation using symbols to connect C and T.   .... 1 mark

c A tank of capacity **5000 l** is to be filled at the same rate as before.
How long, in hours and minutes, does it take to fill the tank?

.... hours .... minutes   2 marks

108

PRACTICE QUESTIONS

4 Janet exchanges £550 for 880 euro.

  a Work out the exchange rate.

  £1 = ... euro   1 mark

  b While on holiday Janet buys a leather bag for 26.40 euro.
  Work out how much this is in pounds.

  £ ....   2 marks

5 The travel graph represents the journeys of two cyclists, Adam and Belinda.
Adam is travelling from Crook to Debens.
Belinda is travelling from Debens to Crook.

  a How many minutes after Belinda did Adam start his journey?

  .... minutes   1 mark

  b Describe briefly what happened at point X.

  1 mark

  c Work out the average speed for Adam's entire journey, in kilometres per hour.

  .... km/h   1 mark

  d Work out Belinda's average speed using only the amount of time during which she was actually moving.

  .... km/h   1 mark

  e Between which two times was Adam's speed the fastest?

  .... and ....   1 mark

109

WHAT YOU NEED TO KNOW

10 Perimeter, area and volume

The world record for the number of people in a phone box is 23.
This is a lot of people in a very small space!

Levels 3–5

- **The total distance around the outside of a shape is called its perimeter.**
  The perimeter of this shape is
  $5 + 8 + 6 + 10 = 29$ cm

- **Area is measured in squares.**
  The area of this shape is 5 squares.
  If each square is 1 cm by 1 cm then the area is 5 cm^2.

- **You can count squares to estimate an area.**
  Count all the squares which are more than half inside the outline.
  Area = 10 cm^2

- **The area of a rectangle = length × width**
  The area of this rectangle is
  $5 \times 3 = 15$ cm^2

- **Units of area are always square units.**

- **Volume is measured in cubes.**
  The volume of this solid is 7 cubes.
  If each cube is 1 cm by 1 cm by 1 cm then the volume is 7 cm^3.

- **The capacity of a hollow object is the volume of space inside it.**
  Capacity is measured in millilitres and litres. 1 m*l* **is the same as 1 cm^3.**

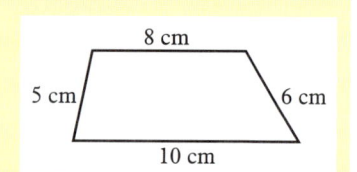

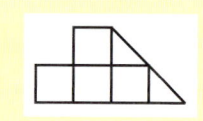

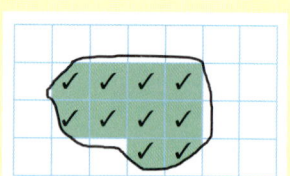

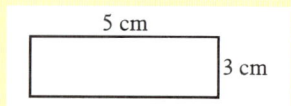

110

WHAT YOU NEED TO KNOW

Level 6

- **The area of a triangle** $= \dfrac{\text{base} \times \text{height}}{2}$

  The area of this triangle is
  $\dfrac{6 \times 5}{2} = \dfrac{30}{2} = 15 \text{ cm}^2$

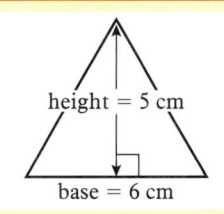

  The base and the height must always be at right angles like these.

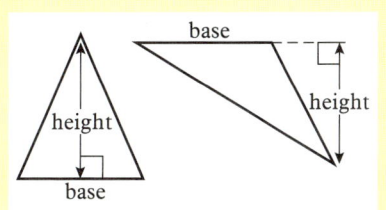

- **Circumference of a circle** $= \pi \times$ **diameter**
  The circumference of this circle is $\pi \times 6 = 18.9$ cm (1 dp)
  The value of π is 3.14159265 …

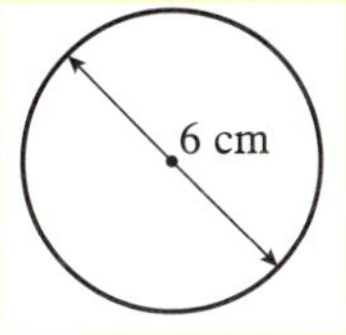

- **Area of a circle** $= \pi \times$ **radius**$^2 = \pi \times$ **radius** $\times$ **radius**
  The area of this circle is $\pi \times 3 \times 3 = 28.3 \text{ cm}^2$ (1 dp)
  The radius is half the diameter.

- **Area of a parallelogram = base × height**
  The area of this parallelogram is
  $10 \times 7 = 70 \text{ cm}^2$

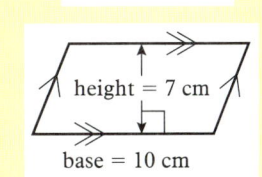

- **The area of a trapezium** $= \tfrac{1}{2}(a + b)h$

  You add the two parallel sides, divide by 2 and then multiply by the height.

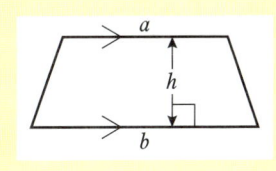

  The area of this trapezium is

  $\tfrac{1}{2}(7 + 5) \times 3 = 6 \times 3 = 18 \text{ cm}^2$

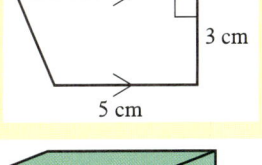

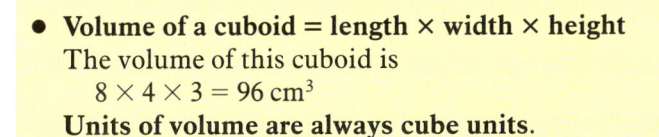

- **Volume of a cuboid = length × width × height**
  The volume of this cuboid is
  $8 \times 4 \times 3 = 96 \text{ cm}^3$
  Units of volume are always cube units.

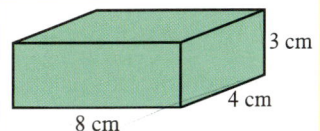

111

TEST YOURSELF

Levels 3–5

1 The perimeter of this shape is

...... + ... + ... = ... cm

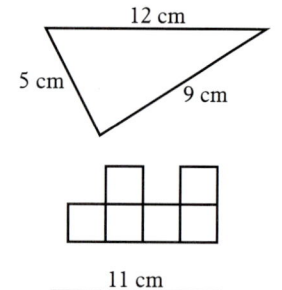

2 The area of this shape is

...... squares

3 The area of this rectangle is

...... cm × ...... cm = ... cm^2

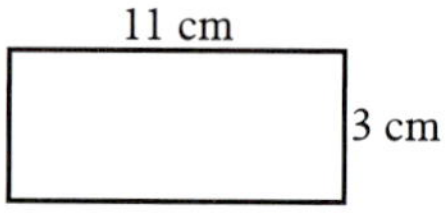

4 The area of this triangle is

$$\frac{\ldots \times \ldots}{\ldots} = \frac{\ldots}{\ldots} = \ldots \text{ cm}^2$$

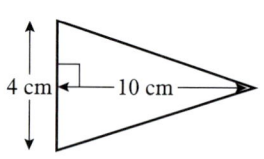

5 The volume of this solid is

...... cubes

Level 6

6 The circumference of this circle is

π × ...... = ... cm (1 dp)

The area of this circle is

π × ...... × ...... = ... cm^2 (1 dp)

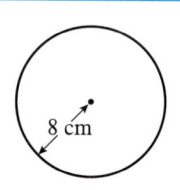

7 The area of this trapezium is

......(...... + ...) × ...... = ... × ...... = ... cm^2

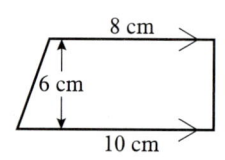

8 The volume of this cuboid is

...... × ...... × ...... = ... cm^3

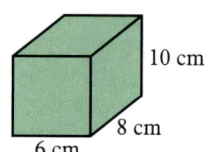

PRACTICE QUESTIONS

Levels 3–5

1 These shapes have been drawn on a 1 cm grid.

Find (i) the perimeter: .... cm   (i) the perimeter: .... cm

(ii) the area: .... cm^2 (ii) the area: .... cm^2

4 marks

2 How many of the **triangular tiles** can fit inside each shape?

Tile

a .... tiles   **b** .... tiles

2 marks

3 This rectangle is made from **six 1 cm square tiles**.

Tile

1 mark

The rectangle has an area of **6 cm^2** and a perimeter of **10 cm**. Draw another shape which has an area of **6 cm^2** but which has a **larger perimeter**.

✓ The tiles must fit together side to side,
✗ not corner to corner.

4 Work out the **area** of each shape.

a 5 cm, 3 cm

b 6 cm, 8 cm

c 4 cm, 7 cm

.... cm^2 .... cm^2 .... cm^2

3 marks

113

PRACTICE QUESTIONS

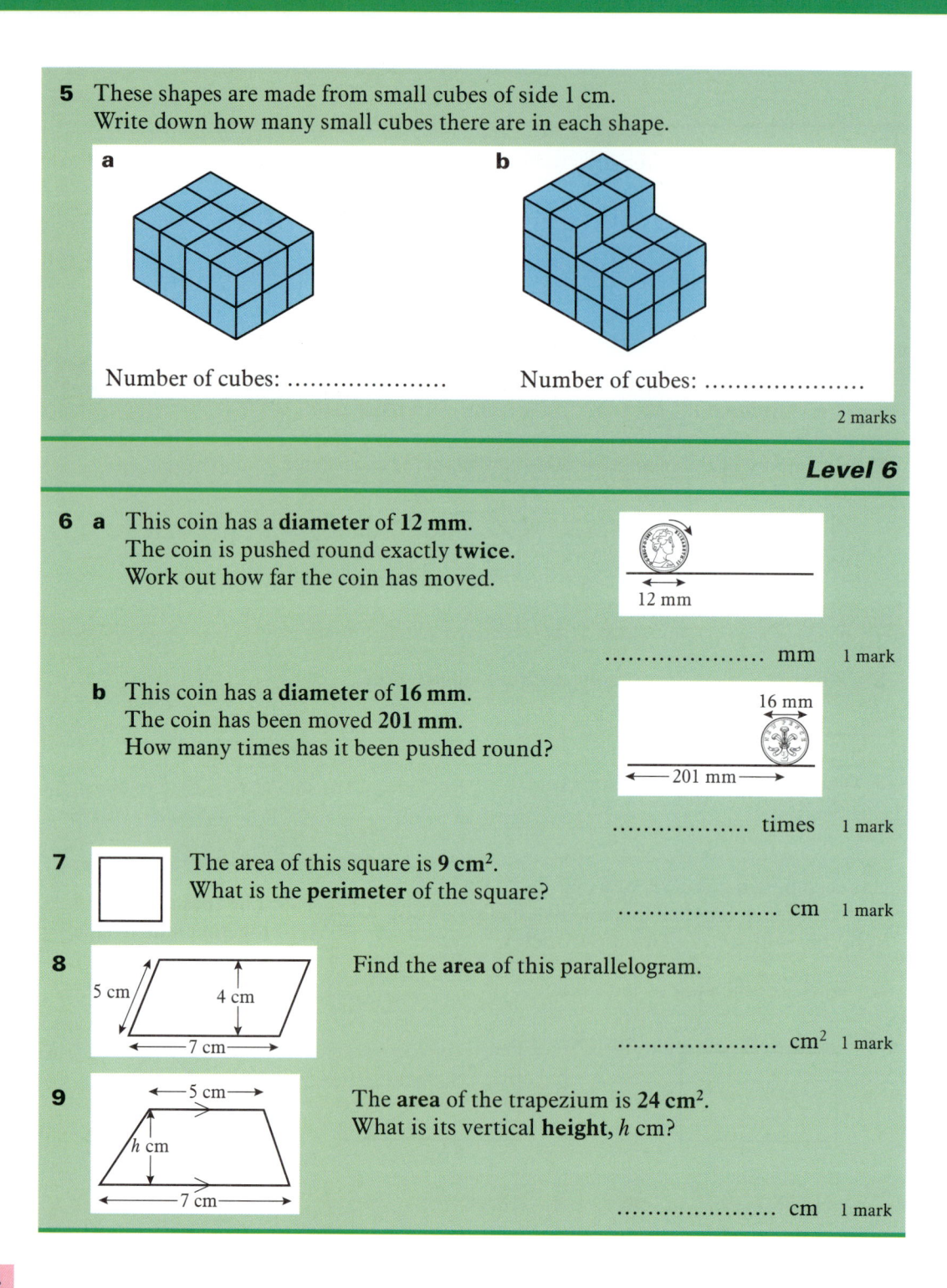

5 These shapes are made from small cubes of side 1 cm.
Write down how many small cubes there are in each shape.

a

b

Number of cubes: ....   Number of cubes: ....

2 marks

Level 6

6 a This coin has a **diameter** of **12 mm**.
The coin is pushed round exactly **twice**.
Work out how far the coin has moved.

.... mm   1 mark

b This coin has a **diameter** of **16 mm**.
The coin has been moved **201 mm**.
How many times has it been pushed round?

.... times   1 mark

7 The area of this square is **9 cm²**.
What is the **perimeter** of the square?

.... cm   1 mark

8 Find the **area** of this parallelogram.

.... cm²   1 mark

9 The **area** of the trapezium is **24 cm²**.
What is its vertical **height**, h cm?

.... cm   1 mark

WHAT YOU NEED TO KNOW

Level 7

- **To find the area of this shape, find the area of the two parts and add them together.**

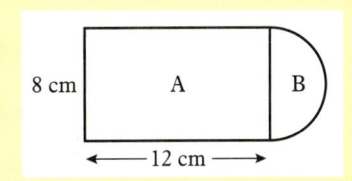

    Area A = 12 × 8 = 96 cm²

    Area B = $\dfrac{\pi \times 4^2}{2}$ = 25.1 cm² (1 dp)

    Total area = 96 + 25.1 = 121.1 cm² (1 dp)

- **To find the area of the red shape you have to subtract the areas.**

    = 12 × 12 − $\dfrac{\pi \times 12^2}{4}$

    = 144 − 113.1

    = 30.9 cm² (1 dp)

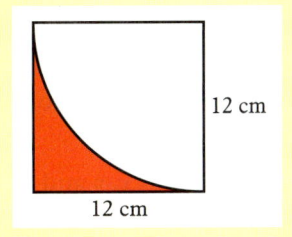

- **Volume of a prism = area of cross section × length**
    A prism is a solid which is exactly the same shape all the way through.

    The volume of this prism is
    15 × 20 = 300 cm³

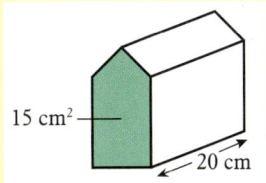

- **Volume of a cylinder = area of cross section × length**
    $= \pi r^2 \times h$

    The volume of this cylinder is
    $\pi \times 8^2 \times 11$ = 2212.0 cm³ (1 dp)

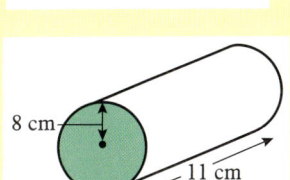

- **Converting area and volume units**

    1 m² = 10 000 cm²   1 m³ = 1 000 000 cm³

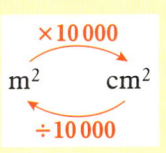

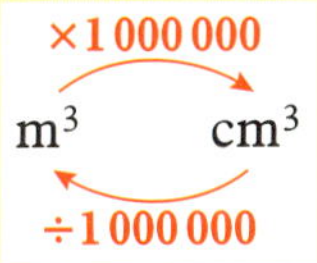

    m² ×10 000 cm²   m³ ×1 000 000 cm³
    m² ÷10 000 cm²   m³ ÷1 000 000 cm³

WHAT YOU NEED TO KNOW

Level 8

- **The dimension of a formula is the number of lengths that are multiplied together.**

- **A constant has no dimension.**

- **Length has one dimension.**
  Formulas for length will only involve constants and lengths.

- **Area has two dimensions.**
  Formulas for area will only involve constants and length × length.

- **Volume has three dimensions.**
  Formulas for volume will only involve constants and length × length × length.

  In these formulas the letters x, y and x are lengths, k is a constant.

  $P = 4xy$ This is constant × length × length
  so P is an area.

  $Q = \pi y^2 z$ This is constant × length² × length
  = constant × length × length × length
  so Q is a volume.

  $R = kx$ This is constant × length
  so R is a length.

- **If a formula has more than one term, each term must have the same dimension.**

  $U = cx + \pi y$ Each term has dimension one
  so U is a length.

  $V = x^2 y - \pi xyz$ Each term has dimension three
  so V is a volume.

  $W = 2y^2 - 4\pi x$ $2y^2$ has dimension two
  $4\pi x$ has dimension one
  so W is an incorrect formula for length or area.

TEST YOURSELF

Level 7

1 The area of this shape is

$$\ldots\ldots + \frac{\ldots\ldots \times \ldots\ldots}{\ldots\ldots}$$

$= \ldots\ldots + \ldots\ldots = \ldots\ldots$ cm² (1 dp)

2 The red area is

$$\ldots\ldots - \ldots\ldots \times \ldots\ldots$$

$= \ldots\ldots - \ldots\ldots = \ldots\ldots$ cm² (1 dp)

3 The volume of this prism is

$$\frac{\ldots\ldots \times \ldots\ldots}{\ldots\ldots} \times 15 = \ldots\ldots \text{ cm}^3$$

4 The volume of this cylinder is

$\ldots\ldots \times \ldots\ldots \times \ldots\ldots$

$= \ldots\ldots$ cm³ (1 dp)

5 The length of this prism is

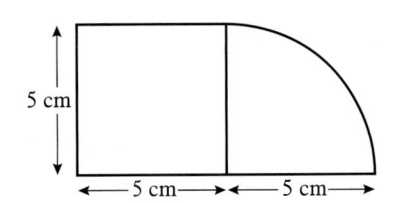

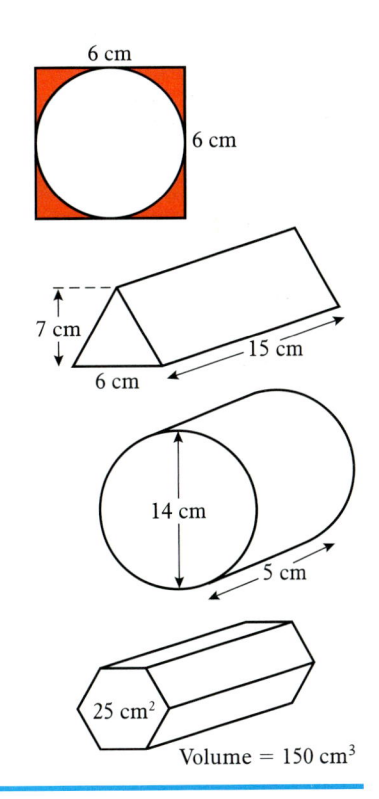

$\ldots\ldots \div \ldots\ldots = \ldots\ldots$ cm

Level 8

6 p, q and r are lengths. c and k are constants.
Write 'length', 'area', 'volume' or 'incorrect' for each of these formulas.

$B = cpq$ .... $D = p^2q + 4pqr$ ....

$C = 2kq + \pi r$ .... $E = 4\pi p^2 + ckr$ ....

117

PRACTICE QUESTIONS

Levels 5–7

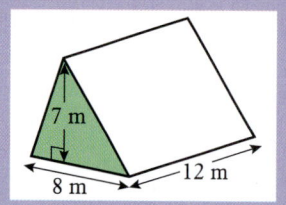

1 **a** Calculate the area of the shaded face.

.... m²   2 marks

b Calculate the volume of the prism.

.... m³   1 mark

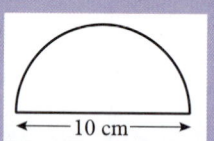

2 For this semicircle find:

a the area   .... cm²   1 mark

b the perimeter   .... cm   1 mark

3 These two triangular prisms have the same cross-sectional area.

What is the height of prism B?

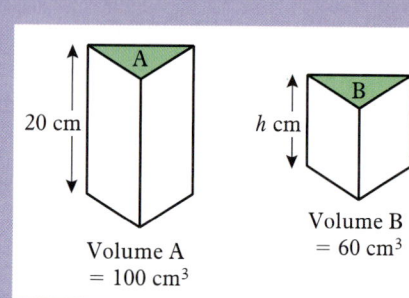

Volume A = 100 cm³

Volume B = 60 cm³

Height: .... cm   2 marks

4 This rectangle has a perimeter of **48 cm**. The length is three times the width. Calculate the **area** of the rectangle.   .... cm²   1 mark

5 A box has internal dimensions as shown. What is the maximum number of cubes, of side 2 cm, that can be stacked inside the cuboid?

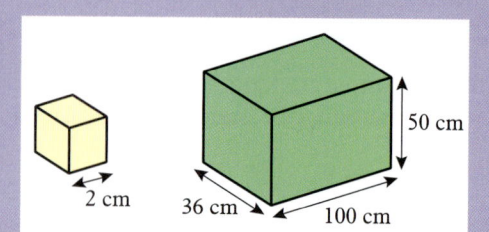

Number: ....   2 marks

PRACTICE QUESTIONS

6 Calculate the area of the shaded region.

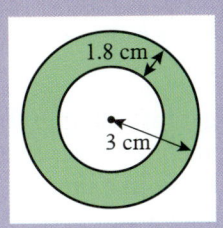

Area: .... cm²   2 marks

7 Calculate the surface area of this triangular prism.

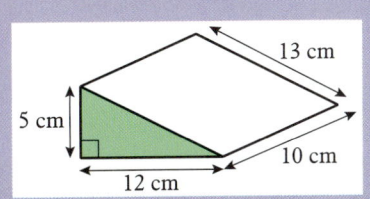

Area: .... cm²   2 marks

Level 8

8 The surface area of a sphere is calculated using the formula $4\pi r^2$, where r is the radius of the sphere.
This hemisphere has a diameter of 8 cm.
Calculate the total surface area of the entire hemisphere.

Area: .... cm²   3 marks

9 Some of the expressions in the table can be used to calculate lengths, areas or volumes of some shapes.
The letters b, h, l and r represent lengths.
π, 2, 3, and 4 are numbers which have no dimension.
Put a tick in a box underneath each expression to show whether the expression represents a length, area or volume.

$4\pi r^2$	lbh	$\frac{1}{2}bh$	$\frac{4}{3}\pi r^3$	$4(r+b)$	$\frac{\pi h^2}{b}$	
						Length
						Area
						Volume

2 marks

WHAT YOU NEED TO KNOW

11 Organising data

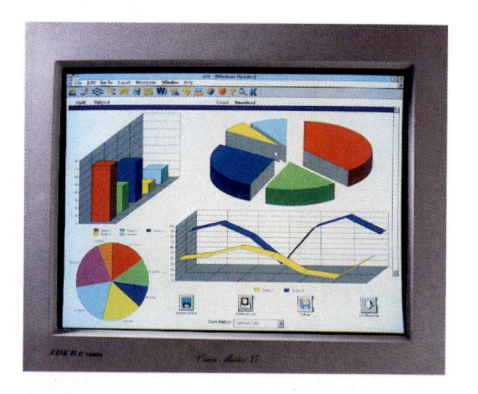

You can get information from lists of data but graphs and charts show it in a more eye-catching way.

Levels 3–5

- **Tally marks are done in groups of five** 卌 . **The total number of tally marks is called the frequency.**

- **Organise data by making a frequency table.**
  This is a frequency table.
  It shows what 30 people have for breakfast.

Breakfast	Tally	Frequency
cereal	卌 卌 II	12
toast	卌 III	8
cooked	IIII	4
drink only	卌 I	6

- **A pictogram is a diagram which uses pictures to show the data. It must always have a key to show what each picture represents.**
  The pictures must all be the same size and in line.

  🥣 represents 4 people
  so 🥣 represents 2 people.

- **A bar-chart uses bars instead of pictures to show the frequency.**
  The bars must all be the same width and the same distance apart. The heights of the bars are the frequencies. You can add up all the heights to find the total number of people.

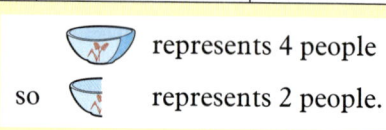

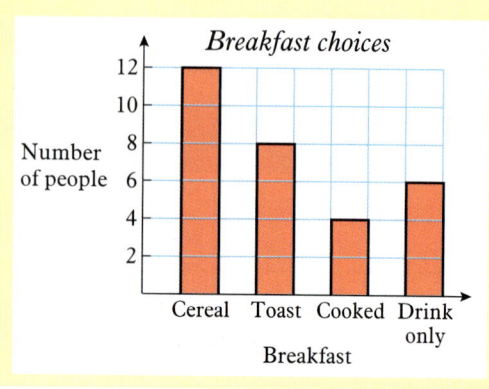

  Breakfast choices

WHAT YOU NEED TO KNOW

- **A stem and leaf diagram uses the data from the bars.**
  This data is shown on the stem and leaf diagram.

  33, 18, 25, 23, 31, 20, 19, 28, 35, 22

  You must always include a key to show you've split the data into stems and leaves.

Stem	Leaf
1	8 9
2	0 2 3 5 8
3	1 3 5

  Key: 2 | 3 means 23

- **You can use a vertical line instead of a bar.**
  This is called a vertical line graph.

- **Instead of a bar-chart or vertical line graph you can draw a frequency polygon.**
  The table shows the KS3 Maths results for 30 pupils.

Level	3	4	5	6
Number of pupils	7	8	10	5

  Plot the points like co-ordinates.
  Join them up with straight lines.

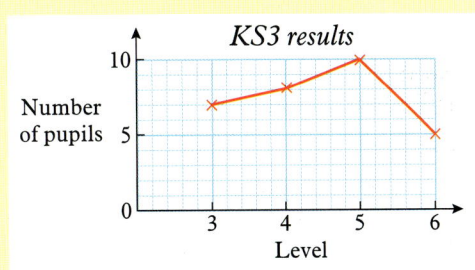

- **When data is put into groups you can still draw bar-charts and frequency polygons.**
  The table shows the percentage test results for 30 pupils.

Score	Number of pupils
50–59	3
60–69	8
70–79	10
80–89	7
90–99	2

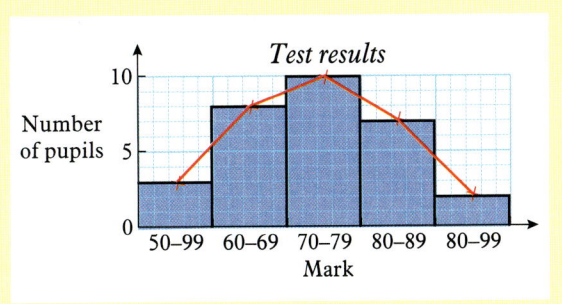

  The bars in the bar-chart now touch.
  The red graph is the frequency polygon.
  Each point is plotted in the middle of the group.

WHAT YOU NEED TO KNOW

- **A conversion graph is a graph that you can use to change from one unit to another.**

    Conversion graphs are always straight lines.
    This graph lets you convert between miles and kilometres.
    To change miles into kilometres you go up to the line and then across.
    To convert 15 miles into km follow the red lines.
    15 miles is 24 km.

    To convert 32 km into miles follow the blue lines.
    Go across and then down.
    32 km is 20 miles.

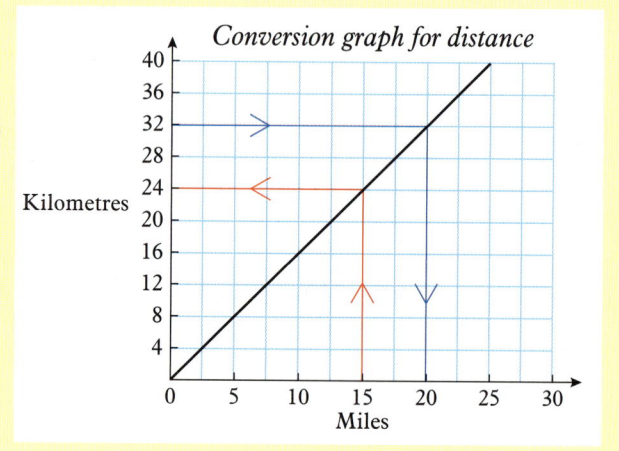

- **Another type of diagram is a pie-chart.**
  The angle of the slice represents the frequency.

    The pie-chart shows that:
    $\frac{1}{2}$ of the chocolates are milk
    This is the same as 50%
    $\frac{1}{4}$ of the chocolates are plain
    This is the same as 25%

    If the total number of chocolates is **40**

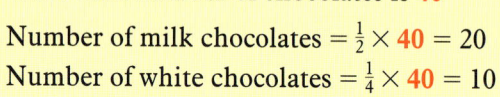

    Number of milk chocolates = $\frac{1}{2} \times$ **40** = 20
    Number of white chocolates = $\frac{1}{4} \times$ **40** = 10

- **You can only estimate the sizes of the slices for some pie-charts.**

    The slice for Fat is a bit more than 25%.
    An estimate is 30%.
    The slice for Carbohydrate is a bit more than 50%.
    An estimate is 60%.

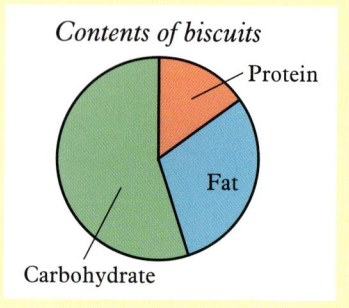

- **All diagrams must be properly labelled and given a title.**

WHAT YOU NEED TO KNOW

Level 6

- **Drawing pie-charts**
  120 students were asked what they want to do when they leave school.
  Here are the results:

Go to university	Go to college	Get a job	Don't know
50	35	25	10

  Draw a pie-chart to show this data.

  First you need to work out the angles for the pie-chart.
  There are 360° to share between the 120 students.
  Each student gets 360° ÷ 120 = **3°**
  Now work out the angle for each slice.

	Number of students	Angle
University	50	50 × **3°** = 150°
College	35	35 × **3°** = 105°
Job	25	25 × **3°** = 75°
Don't know	10	10 × **3°** = 30°

  Check the angles add up to 360°.
  150° + 105° + 75° + 30° = 360° ✔

  You can now draw the pie-chart.
  (1) Draw a circle. Mark the centre.
  Draw a line to the top of the circle.
  Draw the first angle (150°).

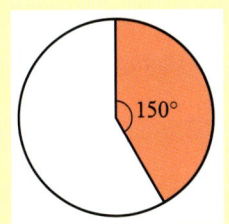

  (2) Measure the next angle
  from the line that you
  have just drawn (105°).

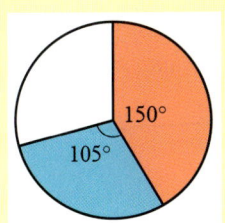

  (3) Carry on until you have drawn all
  the angles.
  Label the pie-chart.
  You need to label each slice like this.
  You also need a title.

  What students hope to do when they leave school

123

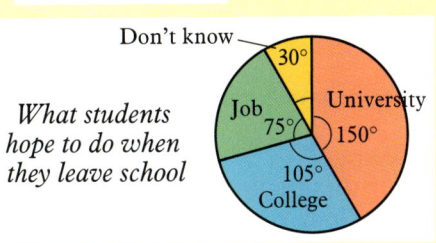

WHAT YOU NEED TO KNOW

- **There are two types of data.**
  When data can only take certain individual values it is called discrete data.
  Number of people is discrete data. You can't have 2.8 people.
  When data can have any value it is called continuous data.
  Length, area, mass, time and temperature are all examples of continuous data.

- **To draw diagrams for continuous data you must label the horizontal axis as a scale.**

Journey times	Number of pupils
5 mins but < 10 mins	5
10 mins but < 15 mins	8
15 mins but < 20 mins	10
20 mins but < 25 mins	7

Time	10:00	12:00	14:00	16:00	18:00	20:00
Temp. (°C)	12	19	20	18	15	10

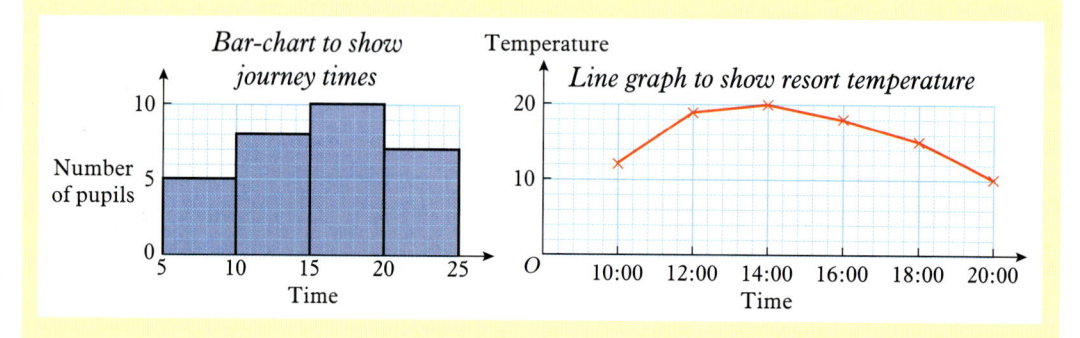

Bar-chart to show journey times

Line graph to show resort temperature

- **Sometimes graphs are used to mislead people.**
  Changing the scale has a big effect.

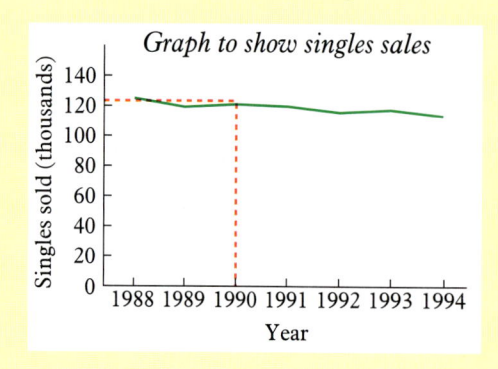

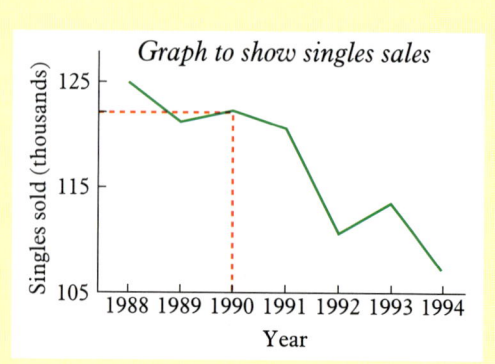

Graph to show singles sales

Graph to show singles sales

The two graphs show exactly the same information.
They look very different because of the scales.

TEST YOURSELF

Levels 3–5

1 This data gives you the number of TVs in 30 families.

2 2 3 1 2 3 1 3 2 1 4 2 1 2 2
1 2 2 1 3 2 3 1 3 2 2 4 2 3 1

a Tally these results.

Number of TVs	Tally	Frequency
1		
2		

b Draw a pictogram.
Use ▢ to represent two TVs.

Number of TVs

| 1 |
| 2 |
| 3 |
| 4 |

c Draw a bar-chart of the data.

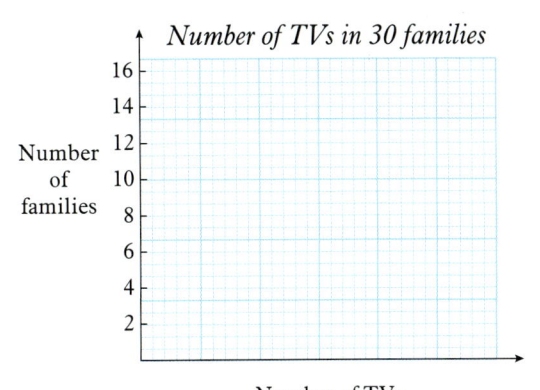

Number of TVs in 30 families

Number of families

2 a The fraction for gas is ………

b ……% use gas.

c About ……% use storage.

50 households are represented.

d …… households use gas.

Types of heating

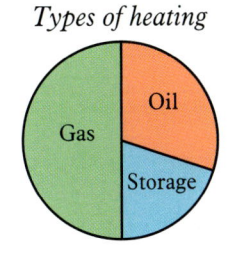

125

TEST YOURSELF

Level 6

3 Complete this pie-chart for these sales of hot drinks

   Coffee   40
   Tea   80
   Chocolate 60

Total number of drinks = ………

Each drink gets 360° ÷ …… = ……°

Angle for coffee = ………

Angle for tea = ………

Angle for chocolate = ………

4 Draw a bar-chart for this data.

Masses of tomatoes (g)	Frequency
0 to less than 10	6
10 to less than 20	12
20 to less than 30	13
30 to less than 40	4

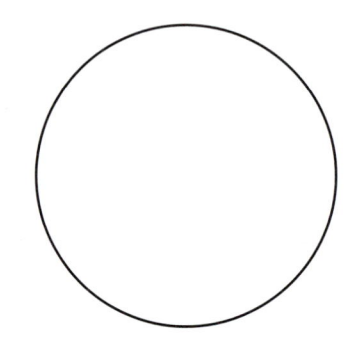

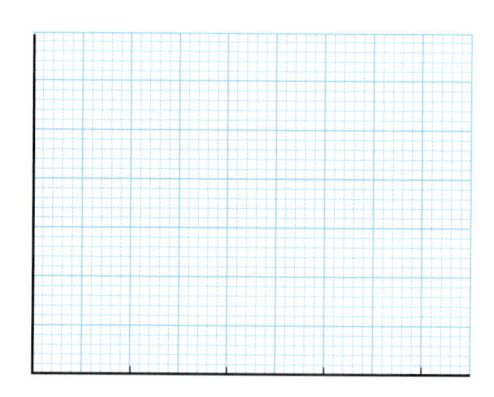

5 This bar-chart shows a company's profits in 1995 and 1996. Why is the chart misleading?

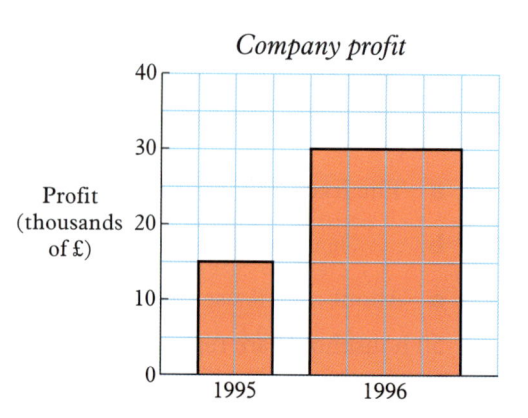

…………………………………………

…………………………………………

…………………………………………

…………………………………………

Company profit

PRACTICE QUESTIONS

Levels 3–5

1 The **pictogram** shows the number of CDs sold on a market stall.
Each symbol represents **4 CDs**.

Number of CDs sold

Monday	◉ ◉ ◉ ◉ ◉ ◉
Tuesday	◉ ◉ ◉ ◉ ◉ ◖
Wednesday	◉ ◉ ◉ ◖
Thursday	
Friday	

a How many CDs were sold on

Monday .... Tuesday .... Wednesday .... 3 marks

b CDs were also sold on **Thursday** and **Friday**.
Show these sales figures on the pictogram as symbols.

9 CDs were sold on Thursday
24 CDs were sold on Friday. 2 marks

c The market was closed for half a day during the week.
On which day was the market closed for half a day? .... 1 mark

The **following** week a **tally-chart** was used to record the sales.

Monday 卌 卌 卌 ll   Tuesday 卌 卌 ll   Wednesday 卌 lll
Thursday 卌 卌 llll   Friday 卌 卌 卌 lll

d Give one reason why the **pictogram** is a good way of recording the sales figures.

....

.... 1 mark

e Give one reason why the **tally-chart** is a good way of recording the sales figures.

....

.... 1 mark

127

PRACTICE QUESTIONS

2 Kim did a survey of **two** types of television programme, on **two** different channels, during a week.

	Channel 1	Channel 3
Films	16	12
Quiz shows	11	13

a How many programmes did Kim record in the week?

.... 1 mark

b What was the **total** number of films shown on **both** channels?

.... 1 mark

c Kim also recorded details of two other types of programme in her survey.

	Mon	Tues	Wed	Thur	Fri
Channel 1	soap, comedy	soap, soap	soap, comedy	soap, comedy	soap, soap, comedy
Channel 3	soap, soap, comedy	soap, soap, soap	soap, soap, comedy	soap, comedy soap, comedy	soap, comedy comedy

Record this data in the table below.

	Channel 1	Channel 3
Soap		
Comedy		

2 marks

3 Malika did a survey of pets in her class. She drew a pie-chart to show her results.

a Which pet was the most popular?

....

b Which two pets were equally popular?

.... and ....

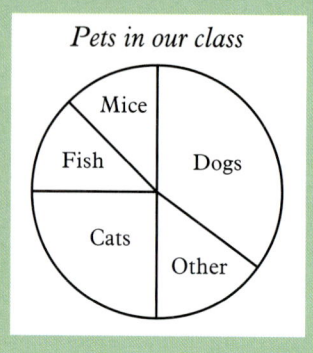

Pets in our class

(pie chart showing: Mice, Fish, Dogs, Cats, Other)

128

PRACTICE QUESTIONS

Malika recorded **40** pets in her survey.

c How many of the pets were cats? ....

d Estimate how many of the pets were dogs? .... 4 marks

Level 6

4 A surveyor in a street asked people to say which one of five different types of holiday they liked best.
The results are shown below.

Ski	|||| |||| |||
Winter abroad	|||| |||| |||| |||| |||| ||||
Summer sun	|||| |||| |||| |||| |||| |||| |||| |||| |||| |||| |||| |||| |||| |||| |||| |||| ||
Touring	|||| |||| |||| |||| ||
UK holiday	|||| |||| |||| |||| |||

a Give a reason why this display of the information is **not** clear.

.... 1 mark

b How many people were questioned? .... 1 mark

c Give a reason why a **pie-chart** rather than a **bar-chart** is a clearer way to show this information?

.... 1 mark

d Draw a pie-chart to show the holiday information.
Label each section of your pie-chart and give it a title.

3 marks

129

WHAT YOU NEED TO KNOW

Level 7

- **Questionnaire design.**

    (1) Questions should not be biased or upset people.

    (2) Questions should be clear and useful to your survey.

    (3) Don't ask questions that give a lot of answers. Structure them so that you can give a choice of possible answers such as

    (a) Yes ☐   No ☐   Don't know ☐

    (b) Agree ☐   Disagree ☐   Don't know ☐

    (c) 0–2 ☐   3–5 ☐   More than 5 ☐

    (d) Strongly agree   1 2 3 4 5   Strongly disagree

- **Frequency polygons can be used to compare two sets of data.**

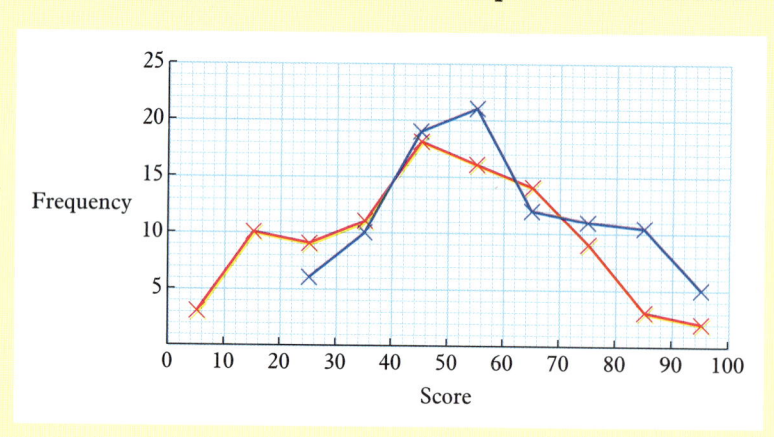

The red and blue frequency polygons show the exam results of 100 pupils at two schools.
The blue results are higher in general than the red results. This is shown by the blue frequency polygon being generally to the right of the red frequency polygon and being above the red polygon for higher marks.
The red results are more spread out than the blue.
This is shown by the width of each polygon.

TEST YOURSELF

Level 7

1 Say what is wrong with each of these questions.

   a Intelligent people watch the TV news. Do you?

      ....

   b How much do you weigh?

      ....

   c What do you think about smoking in restaurants?

      ....

2 50 words are sampled from each of two books and the number of letters in each word counted.

Number of letters	1	2	3	4	5	6	7	8
Number of words, Book A	3	9	14	17	6	1	0	0
Number of words, Book B	4	5	6	10	8	9	5	3

   a Draw a frequency polygon for each book on these axes.

   b Describe the differences between the two books.

      ....

      ....

      ....

      ....

PRACTICE QUESTIONS

Levels 5–7

1 Two snack bars keep a record of their sales for a day. The pie-charts show the sales of certain items.

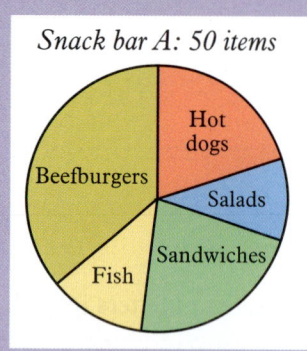

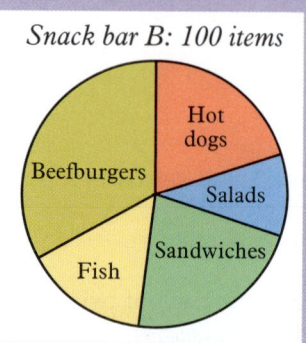

Snack bar A: 50 items   *Snack bar B: 100 items*

a Estimate the **percentage** of people in **Snack bar A** who bought a beefburger.

............%   1 mark

b **100** items were recorded for **Snack bar B**.
Estimate the number of hot dogs that were sold.

....   1 mark

c Explain why the pie-charts do **not** show that more beefburgers were sold in Snack bar A than Snack bar B.

....

.... 1 mark

d This table shows the data collected from a third snack bar.

Snack bar C	
Beefburgers	70
Hot dogs	50
Salads	30
Sandwiches	30
Fish	20

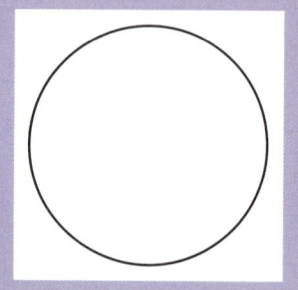

Draw a pie-chart to show the information in the table.
Label each section of your pie-chart and give it a title.

3 marks

PRACTICE QUESTIONS

2 Some pupils wanted to find out how much pupils at their school spend on crisps and chocolate. They wrote a questionnaire.

   a One of their questions is shown below.
'On average, how much do you spend on crisps and chocolate per day?'

☐ 50p or less   ☐ 5–10p   ☐ 10–15p   ☐ 15–20p   ☐ 20–30p   ☐ 30p or over

Write down **three** ways in which these boxes could be improved.

....

....

.... 3 marks

   b One pupil said, 'We can't give a questionnaire to everyone, so let's just give them to our class to complete.'
Explain what is wrong with this method of collecting the data.

....

....

.... 2 marks

3 The frequency polygon shows the number of TV sets sold by shop assistants Ali and Gary in a shop, over a 10 week period.

   a 'It is easy to predict Gary's sales in week 11.' Explain why this statement is true.

.... 1 mark

   b 'Ali has the greatest number of sales overall.' Explain why this statement is false.

.... 1 mark

133

WHAT YOU NEED TO KNOW

12 Averages and spread

A famous politician once said, 'We want everyone to be better than average.'
You need to understand averages better than he did!

Levels 3–5

- **There are three types of average – the mode, the median and the mean.**

- **The mode is the most common or most popular data value.**
  It is sometimes called the modal value.
  The mode is the easiest average to find if there is one.
  The mode of   90 90 95 91 89 91 84 90   is 90

- **The median is the middle value when the data is given in order of size.**
  To find the median of   29 25 0 32 14 22 14

  (1) Put the numbers in order.   0 14 14 22 25 29 32

  (2) Write down the middle number.   Median = 22

  If the number of data values is even you will have two middle numbers. Add these two numbers together and divide by 2 to find the median.
  To find the median of   95 91 89 84 90 88

  (1) Put the numbers in order.   84 88 89 90 91 95

  (2) Add the 2 middle numbers and divide by 2.   Median = $\dfrac{89 + 90}{2}$ = 89.5

WHAT YOU NEED TO KNOW

- **To find the mean of a set of data**

    (1) Find the total of all the data values.

    (2) Divide the total by the number of data values.

    To find the mean of   88 90 79 94 86 91

    (1) Add them all up. $88 + 90 + 79 + 94 + 86 + 91 = 528$

    (2) Divide by 6 $528 \div 6 = 88$

    The mean is 88

- **The range of a set of data is the biggest value take away the smallest value.**

    The range of   72 **43** 62 57 **75** 63   is $75 - 43 = 32$

    The range is a single number not 43 to 75 or 43−75 which you might see in a newspaper.

    The range tells you how spread out the data is.

- **To compare two sets of data you need**

    (1) A measure of average. This will usually be the mean.

    (2) The range.

    Here are the runs scored by two cricketers in their last six innings.

Ian	44	73	39	60	68	40
Gavin	120	7	84	26	9	90

    Ian's mean is 54 runs. His range is 34 runs.

    Gavin's mean is 56 runs. His range is 113 runs.

    The means are very similar but Gavin's range is much bigger. The bigger range shows that Gavin may score a lot of runs but he may score very few. Ian's smaller range means that he is more consistent.

WHAT YOU NEED TO KNOW

Level 6

- **You can find averages from frequency tables.**

  The frequency table shows the number of pit stops in a race.

Number of stops	Frequency	Stops × Frequency
1	2	1 × 2 = 2
2	11	2 × 11 = 22
3	10	3 × 10 = 30
4	1	4 × 1 = 4
Total = 24		Total = 58

  This row shows that 10 drivers stopped 3 times → 3

  To find the mean add the red column to the table.

  $$\text{Mean} = \frac{58}{24} = 2.4 \text{ (1 dp)}$$

  The mode is the number of stops with the highest frequency.

  Mode = 2

  The median is the middle value. The 24 values are

  1 1 2 2 2 2 2 2 2 2 2 2 3 3 3 3 3 3 3 3 3 3 3 4

  $$\text{Median} = \frac{2+2}{2} = 2$$

- **A scatter graph is a diagram that is used to see if there is a connection between two sets of data.**

  Plot pairs of values like co-ordinates and look for a pattern.

  Weight vs Height: As the height increases so does the weight. This is positive correlation.

  Value vs Age: As the age increases the value decreases. This is negative correlation.

  Height vs Test score: There is no connection between height and test score. This is zero correlation or no correlation.

- **Correlation can be strong or weak.**

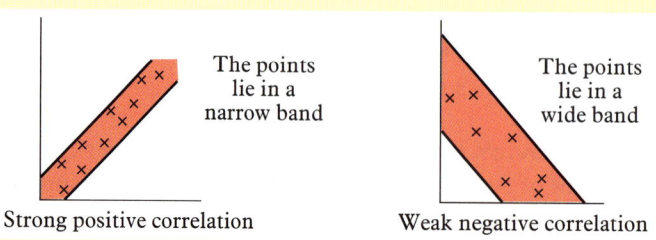

  The points lie in a narrow band — Strong positive correlation

  The points lie in a wide band — Weak negative correlation

TEST YOURSELF

Levels 3–5

1 24   18   15   21   19   18

    a The mode is ………

    b …………………………

        The median is $\dfrac{\ldots\ldots + \ldots\ldots}{2}$ = ………

    c Total = ………

        Mean = ……… ÷ ………

              = ……… (1 dp)

    d Range = ……… − ……… = ………

2 These are the means and ranges for the times that it takes two taxi companies to get a taxi to a customer.

You have to leave to catch a train in 16 mins. Which company would you call? Explain your answer.

	Mean (min)	Range (min)
Terry's Taxis	12.3	13
Chris' Cabs	12.4	4

……

……

……

Level 6

3 This table shows the age and sale value of 10 cars.

Age (years)	3	6	5	4	2	6	5	3	6	5
Value (thousands of £)	6.0	2.5	2.9	3.5	6.5	2.3	3.1	5.8	2.1	3.3

    a Draw a scatter graph.

    b The scatter graph shows

        ………………… correlation

        As ………………… increases

        …………………………………

        …………………………………

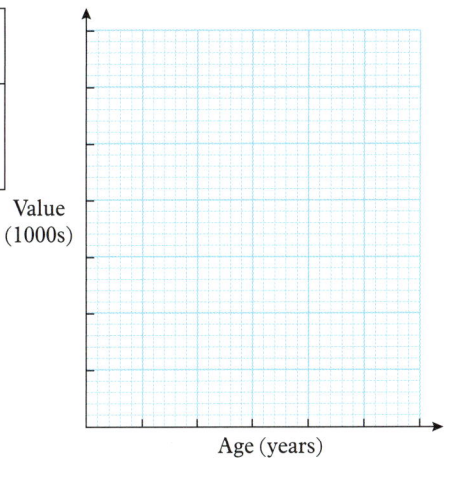

Value (1000s)

Age (years)

PRACTICE QUESTIONS

Levels 3–5

1 3 4 6 2 1 2 3 5 2 5

    a Write down the **mode** of this set of numbers.

    .... 1 mark

    b Write down the **range** of this set of numbers.

    .... 1 mark

    c Write down the **median** of this set of numbers.

    .... 1 mark

    d Write down the **mean** of this set of numbers.

    .... 1 mark

2 Over three games Jerry needs to score an average of **12** points to win.
At his first two attempts he has scored **8** points and **15** points.
What is the **minimum** number of points he must score in his final game to win?

    .... 2 marks

3 Helen conducts a survey to find the number of pets kept in the homes of pupils. She draws a graph of the results of her survey.

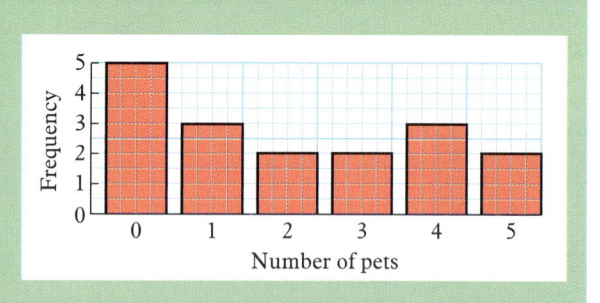

    a Use the graph to calculate the mean number of pets per pupil.

    .... 1 mark

    b What is the modal number of pets?

    .... 1 mark

    c Find the median number of pets.

    .... 1 mark

PRACTICE QUESTIONS

Level 6

4 A teacher records the number of errors on each test paper marked. The results are recorded in the table.

Errors	1	2	3	4	5
Frequency	5	6	7	5	3

a Calculate the **mean** number of errors per pupil.
Show your working.

.... 2 marks

b Find the **median** number of errors per pupil.

.... 2 marks

c Explain why the **mean** and the **median** give different values.

....

.... 1 mark

5 The owner of a cafe keeps a record of the number of cups of hot soup, and the number of cups of coffee that he sells per day, and the temperature on that day.

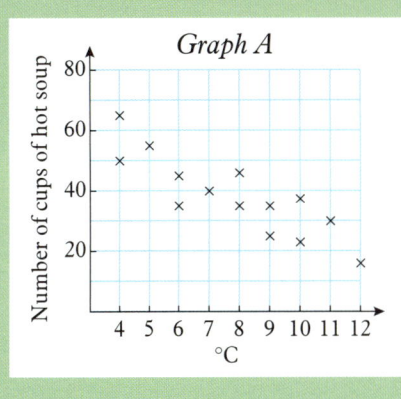

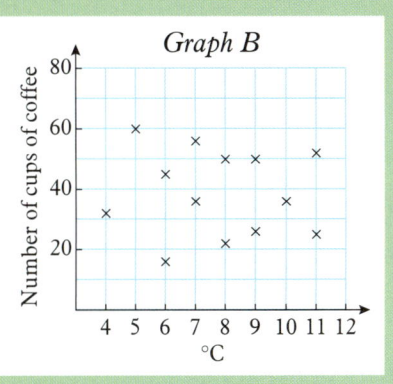

a What does **graph A** show about the relationship between the number of cups of hot soup sold, and the temperature on each day.

.... 1 mark

b What does **graph B** show about the relationship between the number of cups of coffee sold, and the temperature on each day.

.... 1 mark

WHAT YOU NEED TO KNOW

Level 7

- **When a scatter graph shows correlation the points will look as though they lie around a straight line. This line is called the line of best fit. You can use this line to estimate values.**

    This scatter graph shows weights and heights of 6-month-old babies.

    The red line is the line of best fit. You draw this with a ruler. Try to have about the same number of points on each side of the line.

    To estimate the weight for a height of 65 cm draw the blue lines and read off the answer **6.8** kg.

    To estimate the height for a weight of 6 kg draw the green lines and read off the answer **62** cm.

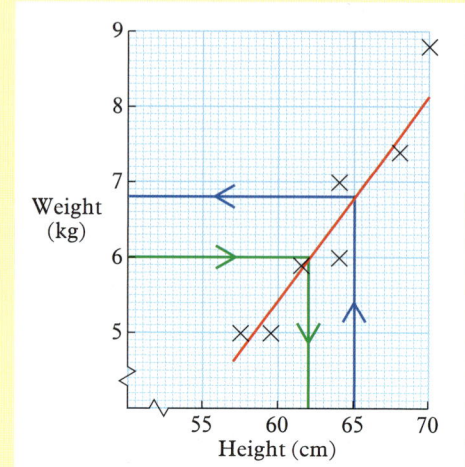

- **Finding mean, mode, median, and range for grouped data.**

    Here are the results of a test.

Mark	31–40	41–50	51–60	61–70	71–80	81–90	91–100
Number of pupils	5	14	28	35	24	16	8

    Look at the first column of data.

    You can see that 5 pupils scored between 31 and 40 but you do not know exactly what each of them scored.

    To work out an **estimate** for the mean you have to assume that all 5 of them scored the middle mark of the group.

    This middle mark is $\frac{31 + 40}{2} = 35.5$

    You can work out all the mid-points and show these in a new table.

Mark	35.5	45.5	55.5	65.5	75.5	85.5	95.5
Number of pupils	5	14	28	35	24	16	8

WHAT YOU NEED TO KNOW

You can now **estimate** the mean as

$$\frac{35.5 \times 5 + 45.5 \times 14 + 55.5 \times 28 + 65.5 \times 35 + 75.5 \times 24 + 85.5 \times 16 + 95.5 \times 8}{5 + 14 + 28 + 35 + 24 + 16 + 8}$$

$$= \frac{8605}{130} = 66.2 \text{ marks (1 dp)}.$$

- **When data is grouped you cannot tell which data value is the most common. You can only say which group has the highest frequency. This group is called the modal group.**
  For the test results the modal group is 61–70 marks.

- **When data is grouped you can only say which group the median is in.**

  For the 130 test results the median would be the $\frac{65^{\text{th}} + 66^{\text{th}}}{2}$ data value.

  The 65th and 66th values lie in the group 61–70 because the first 3 groups contain 47 data values and the first 4 groups contain 82 data values.

  The median is in the 61–70 group.

- **An estimate for the range is the biggest possible value take away the smallest possible value.**
  For the test marks an **estimate** for the range is

  $100 - 31 = 69$ marks.

- **Assumed mean**
  You can find a mean quickly by guessing the answer first. This guess is called the assumed mean.

  Look at this data.   307, 325, 315, 309, 322, 318
  You might guess the mean is 315.
  Now work out the differences between each
  value and this assumed mean.   −8, 10, 0, −6, 7, 3
  Add these differences.   −8 + 10 + 0 − 6 + 7 + 3 = 6
  Divide by the number of data values.   6 ÷ 6 = 1
  Add this amount to the assumed mean   mean = 315 + 1
  to get the true mean.   = 316

WHAT YOU NEED TO KNOW

Level 8

- **Cumulative frequency is a running total.**
  This table shows the lifetimes, in hours, of 375 light bulbs.

The red part of the table is a cumulative frequency table. The red lifetime column shows the upper end of each class.

Lifetime	Frequency	Lifetime	Cumulative frequency
201–400	56	400 or less	56
401–600	124	600 or less	180
601–800	101	800 or less	281
801–1000	63	1000 or less	344
1001–1200	31	1200 or less	375

It is often useful to draw a curve from a cumulative frequency table.

The graph is called a **cumulative frequency curve**.
It allows you to estimate cumulative frequencies for points that are not at the ends of the groups.

A cumulative frequency curve is drawn with the values on the horizontal axis and the cumulative frequency on the vertical axis.
The points are always plotted at the **end** of each range.
The points are joined with a smooth curve.

If you are asked to draw a cumulative frequency diagram you can join the points with straight lines instead of a curve.

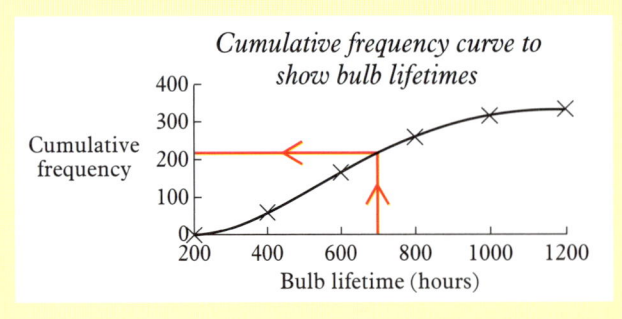

Cumulative frequency curve to show bulb lifetimes

Notice how the curve is joined back to the beginning of the first range in the original table. This is the point (200, 0).

To estimate the number of bulbs that lasted less than 700 hours:
(1) Draw the red line up from 700 on the horizontal axis to the curve and then across to the vertical axis.
(2) Read the value from this axis.
In this example this is approximately 220 bulbs.

142

WHAT YOU NEED TO KNOW

- **The median is the middle data value.**

  To get an estimate of the median:
  (1) Find half the total frequency on the cumulative frequency axis.
  (2) Draw a line across to the curve.
  (3) Draw down to the horizontal axis.
  (4) Read off the estimate of the median.

- **The lower quartile is the value one quarter of the way through the data values.**

  To find the lower quartile:
  (1) Find one quarter of the total frequency on the cumulative frequency axis.
  (2) Draw lines as you did for the median.

- **The upper quartile is the value three quarters of the way through the data.**

- **The interquartile range is the difference between the upper quartile and the lower quartile. This tells you how spread out the central half of the data is.**

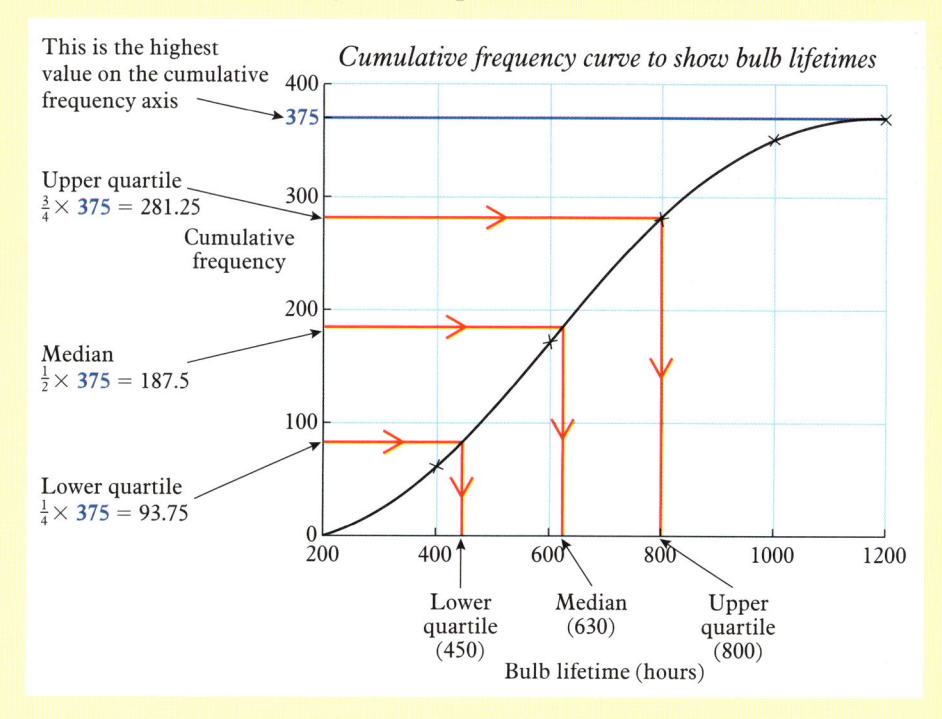

Cumulative frequency curve to show bulb lifetimes

This is the highest value on the cumulative frequency axis → 375

Upper quartile
$\frac{3}{4} \times 375 = 281.25$

Median
$\frac{1}{2} \times 375 = 187.5$

Lower quartile
$\frac{1}{4} \times 375 = 93.75$

Lower quartile (450)   Median (630)   Upper quartile (800)
Bulb lifetime (hours)

Median = 630   Lower quartile = 450   Upper quartile = 800
Interquartile range = 800 − 450 = 350

- **You can use the median and the interquartile range to compare two sets of data.**

TEST YOURSELF

Level 7

1 The table shows the wages of people who work in a car factory.

Wages, £000s	1–10	11–20	21–30	31–100
Wages, mid-points	……	……	……	……
Number of people	25	39	10	6

a Fill in the mid-points in the table.

b Estimate of mean wage = $\dfrac{\ldots \times \ldots + \ldots \times \ldots + \ldots \times \ldots + \ldots \times \ldots}{\ldots + \ldots + \ldots + \ldots}$

  = $\dfrac{\ldots}{\ldots}$ = …… to nearest £1000

c The modal group is ………………

Level 8

2 The cumulative frequency diagrams show the waiting times at a clinic and a surgery.

                  Surgery Clinic

a Median   ……   ……

b Lower quartile   ……   ……

c Upper quartile   ……   ……

d Interquartile range   ……   ……

e Compare the waiting times for the clinic and surgery.

………………………………………………………………………………………………

………………………………………………………………………………………………

………………………………………………………………………………………………

PRACTICE QUESTIONS

Levels 5–7

1 Barry is doing a geography project.
He records on a graph the number of hours of clear sky on certain days, and the temperature at a certain time on these days.

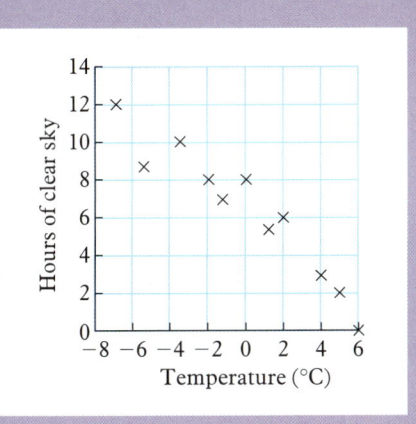

a What does the graph show about the relationship between the hours of clear sky and the temperature in °C?

....

....

....

.... *1 mark*

b Draw a **line of best fit** on the scatter diagram. *2 marks*

Use your line to find an estimate of the hours of clear sky when the temperature is **2°C**.

.... hours *1 mark*

Use your line to find an estimate of the temperature on a day when there are **4 hours of clear sky**.

............°C *1 mark*

145

PRACTICE QUESTIONS

2 Kim counts the number of paces between each of the houses on her paper round. She groups and records the information in a table.

Number of paces	Midpoint (x)	Frequency (f)	fx
30–39	34.5	3	103.5
40–49	44.5	8	
50–59	54.5	11	
60–69	64.5	9	
70–79	74.5	13	
80–89	84.5	6	
Totals		50	

a Calculate an estimate of the **mean** number of paces per house.

.... 2 marks

b Write down the group within which the **median** will lie.

.... 1 mark

PRACTICE QUESTIONS

Level 8

3 The cumulative frequency graph shows the distribution of ages for 200 people in a village.

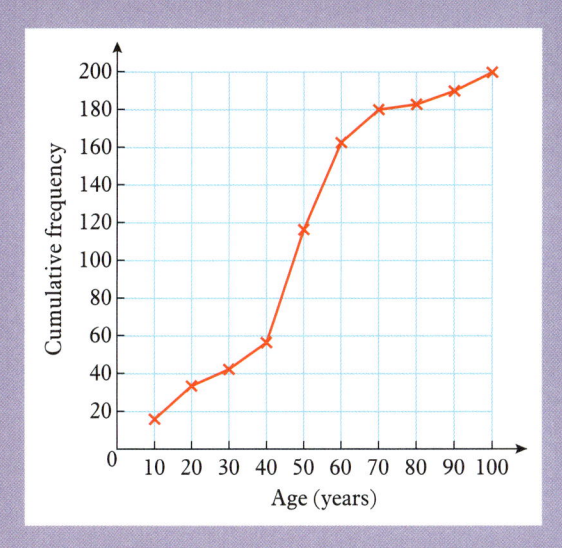

a Use the graph to estimate the **median** age of the people in the village.

.... years   1 mark

b Use the graph to estimate the **interquartile range** of the ages.

.... years   2 marks

c Estimate the **percentage** of the village population who are under 50 years of age.

.... %   1 mark

147

WHAT YOU NEED TO KNOW

13 Probability

People would like to be certain about whether a hurricane is coming their way. Forecasts can only tell them how likely it is. They use probabilities.

Levels 3–5

- **Probability tells you how likely something is to happen.**
- **You can show probabilities on a scale.**

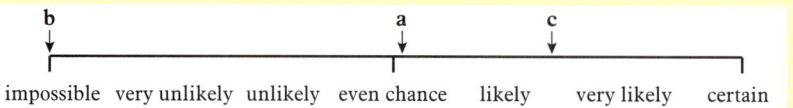

b			a		c	
impossible	very unlikely	unlikely	even chance	likely	very likely	certain

  The scale shows the probabilities that
  (a) a newly born baby will be a girl
  (b) you will live to be 300
  (c) the next person to come into the room will be right handed.

- **You can use probability to decide whether something is fair or not.**
  This spinner is being used by two friends to see who eats the last Rolo.
  One person eats it if the spinner lands on blue, the other person eats it if the spinner lands on yellow. 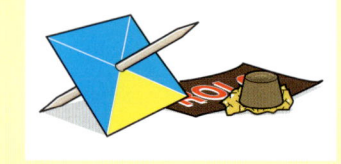
  This is not fair. There is a greater chance that the spinner will land on blue. If you spin it four times you would expect to get three blues and one yellow.

- **Probability is written as a number between 0 and 1.**
  You can only use fractions, decimals, or percentages to show probabilities.

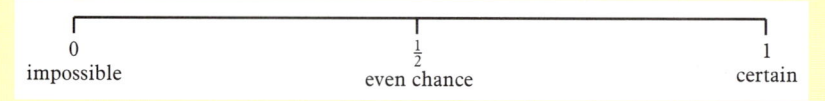

0	$\frac{1}{2}$	1
impossible	even chance	certain

WHAT YOU NEED TO KNOW

- **Events are equally likely if they have the same chance of happening.**

    You can use equally likely events to work out probabilities.

    100 tickets are sold for a raffle.
    The probability that you will win if you have **1** ticket is $\frac{1}{100}$.

    The probability that you will win if you have **5** tickets is $\frac{5}{100}$ because each of the tickets is equally likely to win.

- **There are three methods for working out probabilities.**

    (1) Use equally likely outcomes.
    The probability of getting a 4 with a fair dice is $\frac{1}{6}$

    (2) Use a survey or do an experiment.
    If the events are not equally likely you need to do a survey or experiment to see which is more likely.

    (3) Look back at data.
    If the events are not equally likely and you can't do an experiment, then you look back at data. To find the probability that it will snow in London on Christmas Day look back at the records to see how likely it is.

Level 6

- **Probabilities must add up to 1.**
    If the probability of rain is 0.4 then the probability of no rain is $1 - 0.4 = 0.6$

- **A sample space is a list of all the possible outcomes.**
    A table showing these is called a **sample space diagram**.

    The sample space diagram for the possible outcomes of using this spinner and coin is

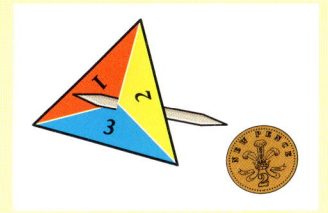

		Spinner		
		1	2	3
Coin	H	H, 1	H, 2	H, 3
	T	T, 1	T, 2	T, 3

    The probability of getting a tail and an odd number $= \frac{2}{6} = \frac{1}{3}$

149

TEST YOURSELF

Levels 3–5

1 Mark these probabilities.
   a The sun will rise tomorrow.
   b Ice will be found on the sun.
   c The sun will shine on the first day of spring.

```
0          0.5          1
```

2 Sue puts these counters into a bag.
   She picks a counter at random.

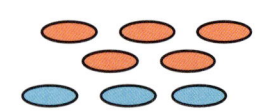

   a What colour is she more likely to get? ....

   b What is the probability that she will get a blue counter? ....

   c Sue picks out a counter and replaces it. She does this 16 times.

   How many red counters does she expect to get? ....

   d She wants to make it equally likely that she will get a red or a blue counter.

   What extra counters does she need to put in the bag? ....

3 Choose the method to find each probability.

   Method 1   Use equally likely outcomes.
   Method 2   Use a survey or do an experiment.
   Method 3   Look back at data.

   a The next car passing will be red.   Method ......

   b You will get an even number when you roll a fair dice.   Method ......

   c There will be an earthquake somewhere in the world today.   Method ......

Level 6

4 a Fill in the sample space diagram.

		\multicolumn{6}{c}{Dice}
		1	2	3	4	5	6
Coin	H				H, 4		
	T						

   b Find the probability of getting a head and an even number. ....

   c Write down the probability of not getting a head and an even number. ....

PRACTICE QUESTIONS

Levels 3–5

1 A box contains cubes which are **five** different colours –
red, green, blue, orange and brown.

There are the same number of cubes of each colour in the box.
Bill takes a cube out of the box without looking.

   a What is the **probability** that Bill will get a **red** or a **blue** cube?

.... 1 mark

   b What is the **probability** that Bill will get a **yellow** cube?

.... 1 mark

   c Draw a cross on the probability scale to show the probability that Bill will get a **blue** or a **brown** cube.

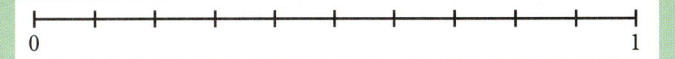

      0 ——————————— 1

.... 1 mark

   d Draw a cross on the probability scale to show the probability that Bill will **not** get a **green** cube.

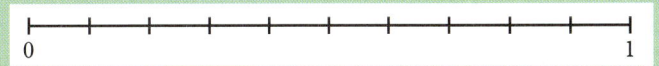

      0 ——————————— 1

.... 1 mark

2 This wheel is spun around.
It will stop with the arrow pointing at a colour.

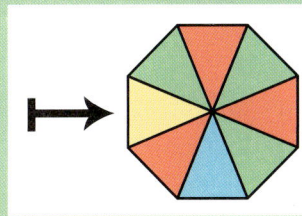

   a Liam says that **red** is more likely than any other colour.
Explain why Liam is **wrong**.

....

.... 1 mark

   b The wheel is spun, and stops with the arrow pointing at **yellow**.
What is the **probability** that the wheel will stop on the **yellow** when it is spun again?

.... 1 mark

151

PRACTICE QUESTIONS

c Colour or shade this spinner so that it will be **certain** that the arrow will land on the **same colour** each time.

1 mark

d Colour or shade this spinner so that there is a **0.5 chance** the spinner will land on **one colour**.

1 mark

3 A coin is thrown and lands on heads **four** times in a row.

  a Draw a cross on the probability scale to show the probability of it landing on heads on the **fifth** throw.

  0 ———————————————— 1

  1 mark

  b **Two** coins are thrown together.
  What is the probability that you will throw two heads together?

  .... 1 mark

  c What is the probability that you will **not** throw two tails with two coins?

  .... 1 mark

Level 6

4 Julie has a bag containing 20 coloured counters.
She takes four red counters out of the bag without looking.

  a Julie says, 'The bag must only contain red counters.'
  Explain why Julie may be **wrong**.

  ....
  ....
  1 mark

  b Julie then takes out two blue and one green counter.
  She says, 'As I have four red, two blue and one green counter then the probability in the future of me picking a red is $\frac{4}{7}$, picking a blue is $\frac{2}{7}$, and a green $\frac{1}{7}$.'
  Explain why Julie is **wrong**.

  ....
  ....
  ....
  1 mark

WHAT YOU NEED TO KNOW

Level 7

- **The frequency of an event is the number of times that it happens.**

- **The relative frequency of an event** $= \dfrac{\text{frequency of the event}}{\text{total frequency}}$

    The relative frequency gives an estimate of the probability.
    This estimate can be improved by increasing the number of times the experiment is repeated.

    A drawing pin is dropped 100 times.
    It lands point up or point down.
    The table shows the results.

    The relative frequency of up $= \dfrac{67}{100}$
    $= 0.67$

Position	Frequency
up	67
down	33

    An estimate of the probability that the drawing pin will land point up is 0.67
    This estimate will get closer to the true probability as the pin is dropped more times.

Level 8

- **Events are mutually exclusive if they cannot happen at the same time.**

    When a coin is thrown it can land showing either a head or a tail. It cannot show a head and a tail at the same time. They are mutually exclusive events.
    When a playing card is chosen from a pack getting an ace and getting a heart are not mutually exclusive. The ace of hearts satisfies both events.

- When combining mutually exclusive events you add the probabilities

    P (diamond) $= \frac{1}{4}$ P (club) $= \frac{1}{4}$
    P (diamond **or** club) $= \frac{1}{4} + \frac{1}{4} = \frac{1}{2}$

- The word '**or**' means **add** the probabilities.

153

WHAT YOU NEED TO KNOW

- **Two events are independent if the outcome of one has no effect on the outcome of the other.**

    When you roll a dice and throw a coin the outcome from the dice has no effect on the outcome of the coin. These are independent events.

- **When combining independent events you multiply the probabilities.**

    P (getting a 4 on a dice) = $\frac{1}{6}$ P (getting head on a coin) = $\frac{1}{2}$
    P (getting a 4 **and** a head) = $\frac{1}{6} \times \frac{1}{2} = \frac{1}{12}$

- The word '**and**' means **multiply** the probabilities.
    Watch for other words like '**both**' being used instead of '**and**'.

- **You can use tree diagrams to show the outcomes of more than one event.**

    This is a tree diagram for throwing two dice when you want to know how many 6s you get.

    P (getting two 6s) = $\frac{1}{6} \times \frac{1}{6} = \frac{1}{36}$

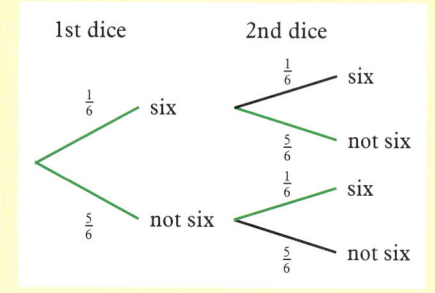

    'Getting two 6s' is the same as 'getting a 6 **and** getting a 6'.
    This is also the same as 'getting a 6 on **both** dice'
    The probability of only getting one 6 is shown by the green routes through the tree.

    P (getting one 6) = P(6 **and** not 6) **or** P (not 6 **and** 6)

    $= \frac{1}{6} \times \frac{5}{6}$ + $\frac{5}{6} \times \frac{1}{6}$

    $= \frac{10}{36}$

    $= \frac{5}{18}$

- **You multiply the probabilities along the branches. You add the probabilities if you use more than one route.**

TEST YOURSELF

Level 7

1. The table shows the colours of cars passing a school.

Colour	Frequency
black	8
red	35
white	28
other	29

   a The total number of cars = ...

   b Relative frequency of black = ...

   c An estimate of the probability that the next car to pass the school will be red = ...

   d How can the estimate in **c** be made more accurate?

   ....

Level 8

2. These two dice are thrown. Find the probability that

   a the red dice will show a 3 or a 5 ....

   b the red dice will show a 3 and the green dice will show an even number ....

   c both dice will show a 6 ....

3. The probability that a pupil has chips for lunch is 0.6

   a Fill in the tree diagram to show the choices of the next two pupils.

   Use the tree diagram to find the probability that

   b the next two pupils both have chips

   ....

   c only one of the next two pupils has chips

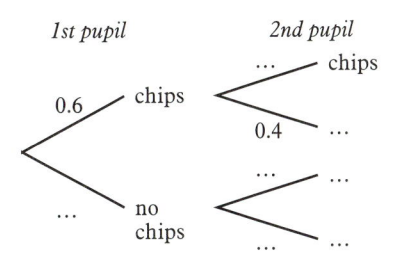

   ....

   1st pupil   *2nd pupil*

   0.6 — chips — ... — chips
                  0.4 — ...
   ... — no chips — ... — ...
                    ... — ...

155

PRACTICE QUESTIONS

Levels 5–7

1 Amanda and Gurjeet each have a spinner.
They **add** the scores on their spinners.

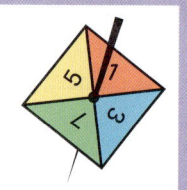

a Complete the table below to show all their possible answers.

+	2	4	6	8
1				
3				
5				
7				

1 mark

b What is the **probability** that their answer is a number **greater than 9**?

.... 1 mark

c What is the **probability** that their answer is a number which is a **square number**?

.... 1 mark

d What is the **probability** that their answer is an **even** number?

.... 1 mark

2 This **biased** cube has coloured faces.
Tony throws the cube **60** times, and records the following results.

Red	Blue	Green	Yellow	Orange	Purple
30	5	6	8	7	4

a Write down an estimate of the probability of throwing a **red** with this cube.

.... 1 mark

PRACTICE QUESTIONS

b How many times would you expect to get a red if you threw the cube **20 times**?

.... 1 mark

c Tony threw the cube **20 times**, but only got a red on **4** occasions.
Can you explain this?

....

.... 1 mark

d How could Tony improve the accuracy of his results?

....

.... 1 mark

Level 8

3 A road traffic survey showed that the probability of a car having to stop at a traffic light is **0.7**
On a particular road there are two sets of traffic lights.

a Calculate the probability that a car will pass through **both** sets of traffic lights without having to stop.

.... 1 mark

b Calculate the probability that a car will be stopped at **only one** of the sets of traffic lights as it passes along the road.

.... 2 marks

c Karen passes through the traffic lights **80 times** in a month. How many times during the month should she expect to be stopped at the **first set** of traffic lights?

.... 1 mark

TEST YOURSELF ANSWERS

1 Whole numbers

Levels 3–5 (p. 8)

1 a 30   b 100
2 a 28   b 54   c 3
3 a 260   b 4100   c 10
4 a 36   b 80   c 10
5 a $\begin{array}{r} 260 \\ +123 \\ \hline 383 \end{array}$ c $\begin{array}{r} 174 \\ -51 \\ \hline 123 \end{array}$

   b $\begin{array}{r} 187 \\ +35 \\ \hline 222 \\ 1\,1 \end{array}$ d $\begin{array}{r} ^45{}^{10}\!\!\not{0}\,9 \\ -46 \\ \hline 463 \end{array}$

6 a $\begin{array}{r} 237 \\ \times25 \\ \hline 1185 \\ 4740 \\ \hline 5925 \\ 1 \end{array}$ b $\begin{array}{r} 23 \\ 14\overline{)32^42} \end{array}$

   $1 \times 14 = 14$
   $2 \times 14 = 28$
   $3 \times 14 = 42$

7 $-6, -4, 0, 5, 12$
8 $3\,°C$

Level 6 (p. 8)

9 a 10   d -5 g -3
   b -11 e 20   h 6
   c 28   f -18 i 50

Level 7 (p. 13)

1 a ÷   b ×   c ×   d ÷
2 a
   (4 . 8 6 × 1 . 6 3) ÷
   (4 . 3 7 + 1 . 9 4) =

   b
   (4 . 2 + 3 . 6 x^2) ÷
   (4 . 9 1 − 2 . 8 7) =

3 a 625   b 9   c 3   d 64

Level 8 (p. 13)

4 a p^9 b m^5 c r^{21} d $\dfrac{1}{4^3} = \dfrac{1}{64}$
5 a 5×10^{-3} b 1.2×10^7
6 a 281 000   b 0.000 000 83
7
   4 . 2 EXP 6 ÷ 2 . 1 EXP
   +/− 4 =

2 Fractions, decimals and percentages

Levels 3–5 (p. 22)

1 a $\begin{array}{r} 2.14 \\ +1.79 \\ \hline 3.93 \\ 1 \end{array}$ c $\begin{array}{r} 2.53 \\ \times24 \\ \hline 10.12 \\ 50.60 \\ \hline 60.72 \end{array}$

   b $\begin{array}{r} 1{}^{1}2.{}^{6}7{}^{1}6 \\ -8.58 \\ \hline 4.18 \end{array}$ d $\begin{array}{r} 1.79 \\ 5\overline{)8.3^{9}4^{5}} \end{array}$

2 a $224 \div 8 = 28$
       $28 \times 5 = £140$
   b $900 \div 100 = 9$ or $72 \div 100 = 0.72$
       $9 \times 72 = 648$ g $0.72 \times 900 = 648$

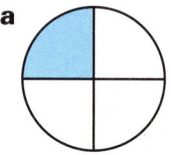

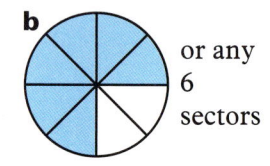

3 a [circle with 1/4 shaded]   b [circle with 3/8 shaded] or any 6 sectors

TEST YOURSELF ANSWERS

Level 6 (p. 22)

4

Fraction	$\frac{1}{2}$	$\frac{1}{4}$	$\frac{1}{5}$	$\frac{3}{4}$	$\frac{3}{5}$	$\frac{81}{100}$
Decimal	0.5	**0.25**	0.2	0.75	**0.6**	0.81
Percentage	50%	25%	**20%**	75%	60%	**81%**

5 a $\dfrac{16}{40} = \dfrac{2}{5}$

  b 35 p as a percentage of 250 p
  $= \dfrac{35}{250} \times 100\% = 14\%$

6 Total number of parts = 9
Value of 1 part = 180 ÷ 9 = £20
Value of each share
$2 \times £20 = £40 \quad 3 \times £20 = £60$
$4 \times £20 = £80$

7 a $\frac{3}{5}$ **c** $\frac{3}{7}$

  b $\frac{8}{12} + \frac{9}{12} = \frac{17}{12} = 1\frac{5}{12}$ **d** $\frac{8}{9} - \frac{6}{9} = \frac{2}{9}$

8 a $\frac{6}{35}$ **b** $\frac{24}{36} = \frac{2}{3}$

  c $\frac{7}{3} \times \frac{3}{2} = \frac{21}{6} = 3\frac{1}{2}$

  d $\frac{3}{4} \times \frac{8}{1} = \frac{24}{4} = 6$

  e $\frac{5}{2} \div \frac{4}{3} = \frac{5}{2} \times \frac{3}{4} = \frac{15}{8} = 1\frac{7}{8}$

9 $\frac{1}{2} = \frac{15}{30} \quad \frac{7}{15} = \frac{14}{30} \quad \frac{3}{5} = \frac{18}{30}$
$\frac{7}{15}, \frac{1}{2}, \frac{3}{5}$

Level 7 (p. 27)

1 a 20% of 650 g = 130 g
New amount = 780 g
  b $\frac{2}{5}$ of £460 = £184
New amount = £644

2 a 15% of 370 cm = 55.5 cm
New amount = 314.5 cm

  b $\frac{3}{8}$ of 2400 kg = 900 kg
New amount = 1500 kg

3 17.5% of 60 = £10.50
Total bill = £70.50

Level 8 (p. 27)

4 Year 1
15% of £8600 = £1290
New value = 8600 − 1290 = £7310
Year 2
15% of £7310 = £1096.50
New value = 7310 − 1096.50 = £6213.50
Year 3
15% of £6213.50 = £932.03
New value = 6213.50 − 932.03
            = £5281.47

5 100% + 7% = 107%
1% = 2568 ÷ 107 = 24
100% = 24 × 100 = 2400

6 $\frac{3}{17}$ is 420 so $\frac{1}{17}$ is 140
Number of people now is
140 × 17 = 2380

3 Estimating

Levels 3–5 (p. 32)

1 a 15 + 3   **d** 7 × 4
   = 18   = 28
  b 26 − 15   **e** 14 ÷ 2 + 7
   = 11   = 7 + 7 = 14
  c 24 − 4   **f** 9 + 24
   = 20   = 33

2 a 2   **c** 140   **e** 100
  b 50   **d** 400

TEST YOURSELF ANSWERS

3 a 34 × 176 = 5984
34 ≈ 30 to the nearest 10
176 ≈ 200 to the nearest 100
30 × 200 = 6000
b 858 ÷ 33 = 26
858 ≈ 900 to the nearest 100
33 ≈ 30 to the nearest 10
900 ÷ 30 = 30

4 a 7.9   **c** 13.4   **e** 4.857
  b 6.32 **d** 29.09 **f** 4.70

5 1 tape costs £11.50 ÷ 8 = £1.44

Level 7 (p. 37)

1 a 4   **c** 30   **e** 7000
  b 0.07   **d** 0.007 **f** 10 000

2 a 500 000   **f** 150
  b 24 000   **g** 300
  c 0.0003   **h** 20
  d 0.000 42 **i** 7000
  e 80   **j** 30 000

3 a $\dfrac{48 \times 35}{6 \times 7}$ **b** $\dfrac{30 \times 42}{5 \times 7}$

  $= \dfrac{48}{6} \times \dfrac{35}{7}$ $= \dfrac{30}{5} \times \dfrac{42}{7}$

  $= 8 \times 5$ $= 6 \times 6$

  $= 40$ $= 36$

Level 8 (p. 37)

4 a $\sqrt{17} \approx \sqrt{16} = 4$ **b** $\sqrt{87} \approx \sqrt{81} = 9$

5 $\sqrt{\dfrac{25 \times 62}{16}} \approx \sqrt{\dfrac{25 \times 64}{16}} = \sqrt{25 \times 4}$
  $= \sqrt{100} = 10$

4 Patterns and sequences

Levels 3–5 (p. 42)

1 a 7, 21, 42   **c** 1, 3, 10, 21
  b 2, 3, 7, 13, 23 **d** 1, 8, 64

2 a 4, 8, 12, 16, 20, 24, 28, 32, 36, 40
  b 9, 18, 27, 36, 45, 54, 63, 72, 81, 90
  c 36

3 a 49   **b** 216   **c** 10   **d** 14

Level 6 (p. 42)

4 a add 5   **c** 3 + 49 × 5 = 248
  b 5   **d** $5n - 2$

Level 7 (p. 46)

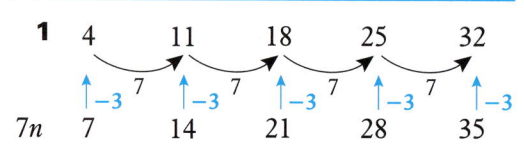

1
4 → 11 → 18 → 25 → 32
(+7, +7, +7, +7)
$7n$: 7, 14, 21, 28, 35 (−3 each)

The formula is $7n - 3$

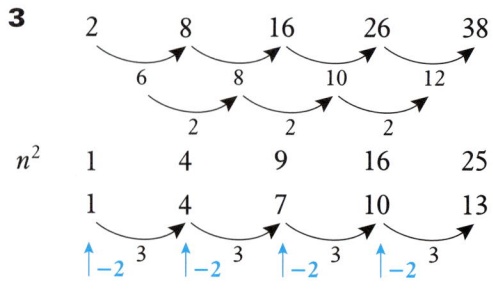

2 a 4 × 1 − 3 = 1
  b 4 × 6 − 3 = 21
  c 4 × 20 − 3 = 77

3
2, 8, 16, 26, 38 (+6, +8, +10, +12; second difference 2)
n^2: 1, 4, 9, 16, 25
1, 4, 7, 10, 13 (−2 difference 3)

The formula is $n^2 + 3n - 2$

4 a 3 − 1 − 2 = 0
  b 3 × 25 − 5 − 2 = 68

160

TEST YOURSELF ANSWERS

5 Formulas, expressions and equations

Levels 3–5 (p. 51)

1 a $c = 25 + 15h$
  b $25 + 15 \times 4 = £85$

2 $P = 2l + 2w$

Level 6 (p. 51)

3 a $5c + 6d$
  b $10q + 35$
  c $3(3p - 4)$
  d $z(z^2 + 4z - 8)$

4 a $x + 7 - 7 = 13 - 7$
      $x = 6$
  b $\dfrac{x}{6} - 3 + 3 = 4 + 3$
      $\dfrac{x}{6} = 7$
      $\dfrac{x}{6} \times 6 = 7 \times 6$
      $x = 42$
  c $5x - x - 28 = x - x - 6$
      $4x - 28 = -6$
      $4x = 22$
      $x = 5.5$
  d $4x - 12 = 28$
      $4x = 40$
      $x = 10$

Level 7 (p. 57)

1 a $2x \leq 8$
      $x \leq 4$
  b $\dfrac{x}{5} > 3$
      $x > 15$

2 0, 1, 2, 3, 4

3 a $x + c^2 = 2r$
      $r = \dfrac{x + c^2}{2}$
  b $\sqrt{p} \leq r - s$
      $r = \sqrt{p} + s$

4 a $2(7q - 4)$
  b $8a(2a^2 - 1)$

Level 8 (p. 57)

5 a $x \geq 0$
  b $y \leq x + 3$
  c $-4 < y \leq 4$

6 a $x^2 - 3x + 4x - 12 = x^2 + x - 12$
  b $2x^2 - 12x - 3x + 18 = 2x^2 - 15x + 18$

6 Functions and graphs

Levels 3–5 (p. 62)

1 A (4, 2) B (1, 4) C (0, 2)

2

Level 6 (p. 62)

3 $y = x + 2$

4 A $x = 2$ C $y = -x$
    B $y = 2x + 2$ D $y = -x - 3$

161

TEST YOURSELF ANSWERS

5 When $x = 0$ When $y = 0$
  $-2y = 6$ $3x = 6$
  $y = -3$ $x = 2$
This gives $(0, -3)$ This gives $(2, 0)$

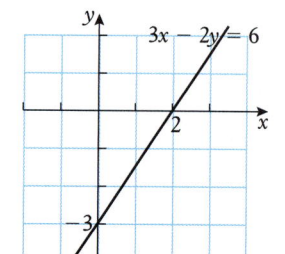

3 Height increases slowly at first but then rate of increase gets bigger. Then constant rate of increase for the last part.

Level 7 (p. 69)

1 a Adding gives $5x = 20$
                       $x = 4$
Put $x = 4$ into (1)
  $2 \times 4 + y = 14$
  $y = 6$
Solution is $x = 4, y = 6$
Check in (2)
  $3 \times 4 - 6 = 12 - 6 = 6$ ✓

b Multiply (1) by 3 and (2) by 2
  $12x + 6y = 30$
  $10x + 6y = 24$
Subtracting $2x = 6$
             $x = 3$
Put $x = 3$ into (1)
  $4 \times 3 + 2y = 10$
  $y = -1$
The solution is $x = 3, y = -1$
Check in (2) $5 \times 3 + 3 \times -1 = 12$ ✓

Level 8 (p. 69)

2 A $y = x^2 - 4x + 7$
    B $y = \dfrac{1}{x - 3}$
    C $y = x^3 - 6x^2 + 11x - 6$
    D $y = 4 - x^2$

7 2D and 3D shapes

Levels 3–5 (p. 77)

1 a obtuse   **b** acute   **c** reflex

2 a W   **c** E
  b SE **d** NE

3 a $70 + 110 = 180$ **110°**
  b $35 + 55 + 90 = 180$ **55°**
  c $80 + 20 + 80 = 180$ **80°**

Level 6 (p. 77)

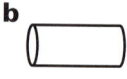

4 a   **b**   **c**

5 a $90 + 90 + 40 + 80 = 300$
     $360 - 300 =$ **60°**
  b $40 + 140 = 180$ **140°**

6 a Angles on straight line
     $130 + 50 = 180$ **50°**
  b Opposite angles **80°**
  c Angles in triangle
     $50 + 80 + 50 = 180$ **50°**

TEST YOURSELF ANSWERS

Level 7 (p. 83)

1 a $x^2 = 8^2 + 13^2$
    $= 233$
    $x = \sqrt{233}$
    $x = 15.3$ (3 sf)

b $12.3^2 = x^2 + 7.8^2$
    $12.3^2 - 7.8^2 = x^2$
    $x^2 = 90.45$
    $x = \sqrt{90.45} = 9.51$ (3 sf)

Level 8 (p. 83)

2 a $\tan 38 = \dfrac{p}{14.1}$

$14.1 \times \tan 38 = p$
$p = 11.0$ cm (3 sf)

b $\cos p = \dfrac{11}{17}$

$p = 49.7°$ (3 sf)

3 $\sin 54 = \dfrac{x}{500}$
$x = 500 \times \sin 54$
$= 405$ miles (3 sf)

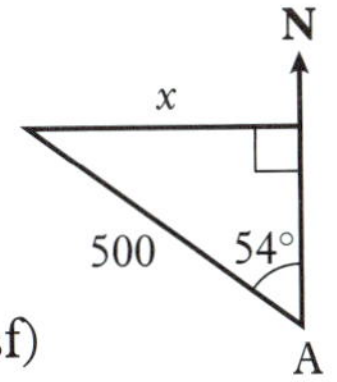

8 Position and movement

Levels 3–5 (p. 90)

1 a order 3
b order 2
c order 4
d order 3

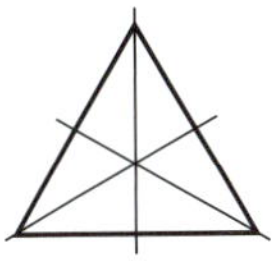

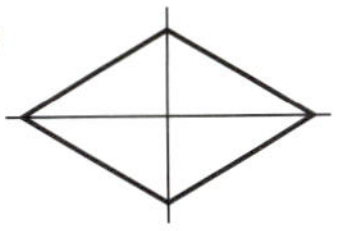

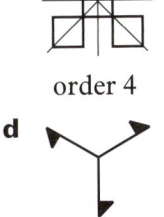

2

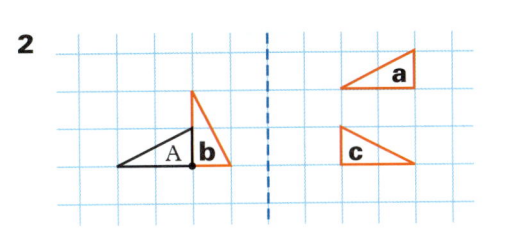

Level 6 (p. 90)

3

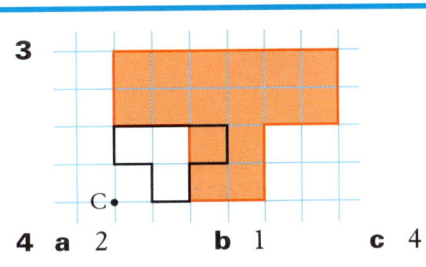

4 a 2   **b** 1   **c** 4

Level 7 (p. 96)

1

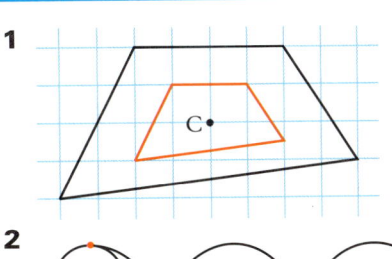

2 a

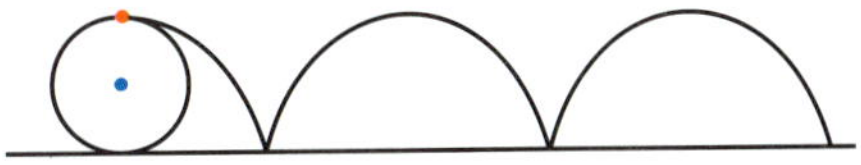

b

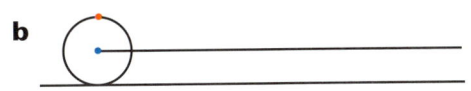

3 The region is shaded.
The boundaries are not included.
Scale: 1 cm to 2 km

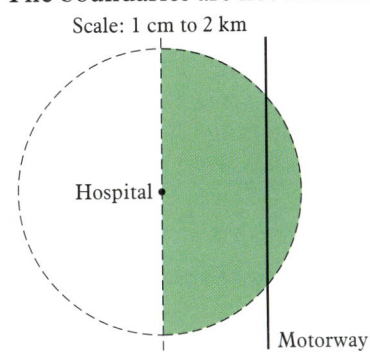

163

TEST YOURSELF ANSWERS

Level 8 (p. 96)

4 a

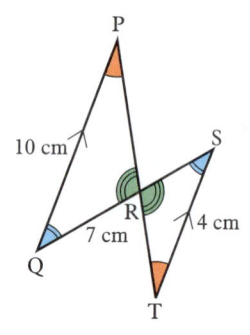

b (1) ST (2) QR
c $10 \div 4 = 2.5$ (or $2\frac{1}{2}$)
d $7 \div 2.5 = 2.8$ cm

9 Units of measurement

Levels 3–5 (p. 103)

1 **a** 100 **e** 8 **i** 4
   b 12   **f** 1000 **j** 48
   c 16   **g** 20 **k** 18
   d 1000 **h** 3000 **l** 2

2 **a** 1.6 **c** 90 **e** 30
   b 600 **d** 4.5 **f** 450

3 **a** 2, 18 **b** 0.1, 5.6 **c** 2, 13

4 **a** 06:50 **c** 1:45 am
   b 11:10 pm **d** 16:36

5 **a** Any in range 3 to 6 m.
   b Any in range 3 to 8 oz.
   c Any in range 2 to 4 in.
   d Any in range 30 to 100 ml.

Level 7 (p. 107)

1 **a** 19.5 cm, 14.5 cm, 5.5 cm
   b 20.5 cm, 15.5 cm, 6.5 cm

2 **a** 60 km   **b** $1\frac{1}{2}$ hours
   c After. The gradient is steeper after his rest.
   Speed before $= 20 \div 1\frac{1}{2} = 13\frac{1}{3}$ km/h
   Speed after $= 40 \div 2 = 20$ km/h
   d $60 \div 5 = 12$ km/h

3 **a** $13 \div 2 = 6.5$ m/s
   b $5 \times (2.5 \times 60) = 750$ m
   c $4 \div 12 = \frac{1}{3}$ h $= 20$ mins
   d $40 \div 1\frac{1}{4} = 32$ mph

4 **a** $50 \div 10 = 5$ g/cm^3
   b $100 \div 50 = 2$ m^3
   c $15 \times 20 = 300$ g
   d $(2.5 \times 1000) \div 100 = 25$ g/cm^3

10 Perimeter, area and volume

Levels 3–5 (p. 112)

1 $12 + 9 + 5 = 26$ cm

2 6 squares

3 $11 \times 3 = 33$ cm^2

4 $\dfrac{4 \times 10}{2} = \dfrac{40}{2} = 20$ cm^2

5 6 cubes

Level 6 (p. 112)

6 $\pi \times 16 = 50.3$ cm (1 dp)
   $\pi \times 8 \times 8 = 201.1$ cm^2 (1 dp)

7 $\frac{1}{2}(8 + 10) \times 6 = 9 \times 6 = 54$ cm^2

8 $6 \times 8 \times 10 = 480$ cm^3

TEST YOURSELF ANSWERS

Level 7 (p. 117)

1 $25 + \dfrac{(\pi \times 5^2)}{4}$
  $= 25 + 19.6349\ldots$
  $= 44.6$ cm^2 (1 dp)

2 $36 - \pi \times 3^2$
  $= 36 - 28.2743\ldots$
  $= 7.7$ cm^2 (1 dp)

3 $\dfrac{6 \times 7}{2} \times 15 = 315$ cm^3

4 $\pi \times 7^2 \times 5 = 769.7$ cm^3 (1 dp)

5 $150 \div 25 = 6$ cm

Level 8 (p. 117)

6 B Area   D Volume
   C Length E Incorrect

11 Organising data

Levels 3–5 (p. 125)

1 a

Number of TVs	Tally	Frequency
1									8
2													13
3								7
4				2

b *Number of TVs*

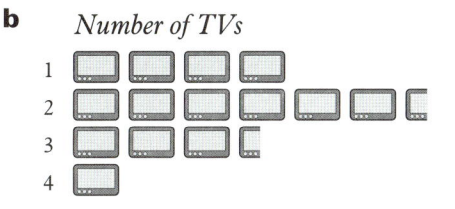

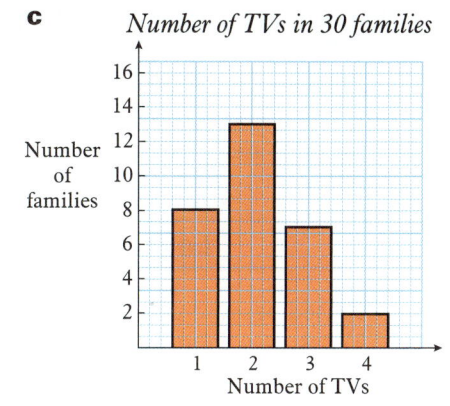

c *Number of TVs in 30 families*

2 a $\frac{1}{2}$ **b** 50% **c** 20% **d** 25

Level 6 (p. 126)

3 Total number of drinks = 180
   Each drink gets $360 \div 180 = 2°$
   Angle for coffee = $80°$
   Angle for tea = $160°$
   Angle for chocolate = $120°$

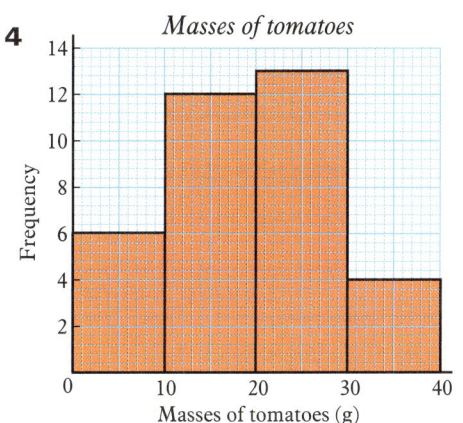

Sales of hot drinks

4 *Masses of tomatoes*

165

TEST YOURSELF ANSWERS

5 Chart is misleading because the bar for 1996 is twice as wide as the 1995 bar. The area is affected by more than the change in profits and the eye will register the area rather than just the height.

Level 7 (p. 131)

1 a Biased because people will answer Yes as they want to appear intelligent.
  b There will be a wide variety of answers – they could well be all different.
  c People will give opinions and these will be impossible to analyse properly.

2 a

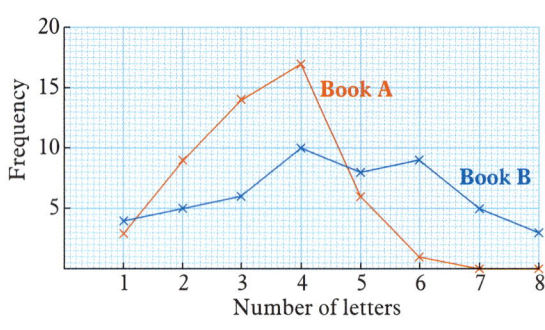

  b There are more shorter words in A than in B. B has a greater range of the number of letters in words.

12 Averages and spread

Levels 3–5 (p. 137)

1 a The mode is 18
  b 15, 18, 18, 19, 21, 24

The median is
$$\frac{18 + 19}{2} = 18.5$$

  c Total = 115
     Mean = 115 ÷ 6 = 19.2 (1 dp)
  d Range = 24 − 15 = 9

2 Chris' Cabs because the times are more consistent and more likely to be less than 16 mins.

Level 6 (p. 137)

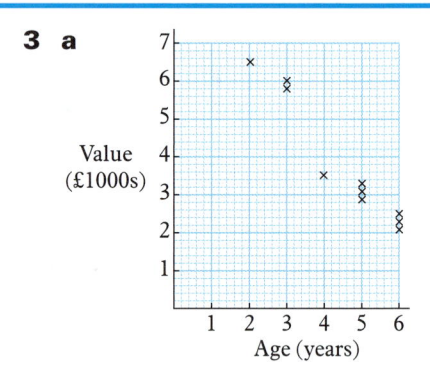

3 a

  b Negative correlation.
     As the age increases the value decreases.

Level 7 (p. 144)

1 a 5.5, 15.5, 25.5, 65.5
     (these are in £1000s)
  b
$$\frac{25 \times 5500 + 39 \times 15\,500 + 10 \times 25\,500 + 6 \times 35\,500}{25 + 39 + 10 + 6}$$
$$= \frac{1\,390\,000}{80} = £17\,375$$

  c The modal group is 11–20

166

TEST YOURSELF ANSWERS

Level 8 (p. 144)

		Surgery	Clinic
a	Median	20	22
b	Lower quartile	14	13
c	Upper quartile	27	32
d	Interquartile range	13	19

e The waiting times at the surgery are generally shorter than those at the clinic. They are also more consistent in length of waiting times.

13 Probability

Levels 3–5 (p. 150)

1
```
    b        c           a
    |--------|-----------|
    0       0.5          1
```

2 a red
   b $\frac{3}{8}$
   c 10
   d 2 blue counters

3 a Method 2
   b Method 1
   c Method 3

Level 6 (p. 150)

4 a

		1	2	3	4	5	6
Coin	H	H,1	H,2	H,3	H,4	H,5	H,6
	T	T,1	T,2	T,3	T,4	T,5	T,6

   (Dice across the top)

   b $\frac{3}{12} = \frac{1}{4}$ c $1 - \frac{1}{4} = \frac{3}{4}$

Level 7 (p. 155)

1 a 100   b $\frac{8}{100} = \frac{2}{25}$
   c $\frac{35}{100} = \frac{7}{20}$ (or 0.35)
   d By surveying more cars

Level 8 (p. 155)

2 a $\frac{2}{6} = \frac{1}{3}$
   b $\frac{1}{6} \times \frac{1}{2} = \frac{1}{12}$
   c $\frac{1}{6} \times \frac{1}{6} = \frac{1}{36}$

3 a

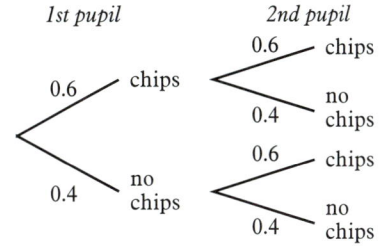

```
         1st pupil              2nd pupil
                          0.6  ─── chips
              0.6  chips
           ╱              0.4  ─── no chips
          ╱
          ╲              0.6  ─── chips
              0.4  no chips
                          0.4  ─── no chips
```

   b $0.6 \times 0.6 = 0.36$
   c $(0.6 \times 0.4) + (0.4 \times 0.6)$
      $= 0.24 + 0.24 = 0.48$

PRACTICE QUESTION ANSWERS

At appropriate stages the marks to be awarded are shown as (✓).

1 Whole numbers

Levels 3–5 (p. 9)

1. **a** 2345 (✓)   **b** 5432 (✓)   **c** 0 (✓)

2. **a** £13 × 25 = 260 (✓) + 65 = £325 (✓)
   25 CDs cost £325

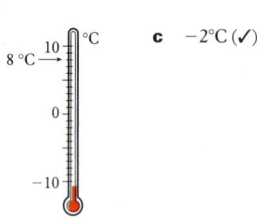

   b $18\overline{)3\ 0^{12}0}$ = 1 6 r12 (✓ *if remainder 12 shown*)
   16 packs of video tapes can be bought with £300 (✓)

3. **a** −3°C (✓)   **b** 8 °C   **c** −2°C (✓)

Level 6 (p. 10)

4. **a** −4 − 5 = −9 (✓)   **b** −4 − −5 = 1 (✓)

5. $T = \dfrac{4 \times 15}{-2} = \dfrac{60}{-2} = -30$ (✓)

Levels 5–7 (p. 14)

1. **a** 821 − 647 = **174** (✓)
   b $a \times b \times 5 = 40$, where $a \times b = 8$ (e.g. 2 × 4) (✓)
   c 1400 ÷ 100 = **14** (✓)
   d 182 ÷ 7 = **26** (✓)
   e $a − b = 31$ (e.g. 32 − 1) (✓)
   f 37 × 9 = **333** (✓)

2. **a** £27 × 34 = 810 (✓) + 108 = £918 (✓)
   34 candles sell for £918

   b $12\overline{)2\ 0^80}$ = 1 6 r8 (✓ *with remainder 8*)
   16 full boxes can be packed (✓)

3. $K = \dfrac{14.2 \times 37.4}{80.4 − 13.29} = \dfrac{531.08}{66.87} = 7.941\,977$ (✓)

4. A   86 850 000 ÷ 347 400 = £250
   B   119 945 000 ÷ 521 500 = £230
   C   98 784 000 ÷ 403 200 = £245
   D   170 977 500 ÷ 670 500 = £255
   a Town B has the lowest tax per person (✓)
   b Town D has the highest tax per person (✓)

Level 8 (p. 15)

5. **a** 1.5×10^{10} (✓)
   b $1.5 \times 10^{10} \times 9.46 \times 10^{12}$ (✓) $= 1.419 \times 10^{23}$ (✓)

6. $10^{600} \div 10^{303}$ (✓) $= 10^{600−303} = 10^{297}$ (✓)

2 Fractions, decimals and percentages

Levels 3–5 (p. 24)

1. **a** 16 ÷ 2 = 8 (✓)   **c** 16 − **a** or 16 − 8 = 8 (✓)
   b $\frac{4}{16} = \frac{1}{4}$ (✓)   **d** 16 − 4 = 12 (✓)

2. To find the quantities needed for 9 people, first divide by 6 and then multiply by 9 (i.e. multiply by 1.5).
   Garlic: 2 ÷ 6 × 9 = 3 cloves
   Chick peas: 4 ÷ 6 × 9 = 6 ounces
   Olive oil: 4 ÷ 6 × 9 = 6 tablespoons
   Paste: 5 ÷ 6 × 9 = $7\frac{1}{2}$ (fluid ounces) (2✓)

3. **a** $\frac{9}{12} = \frac{3}{4}$ (✓)
   b $\frac{3}{4} \times 100 = 75\%$ (✓)
   c $\frac{2}{3} \times 12 = 8$ shaded triangles (✓)

4. **a** $\frac{5}{100} \times 30 = £1.50$ (✓)
   b $\frac{2}{5} \times 4 = £1.60$ (✓)
   c £1.50 (✓)

Level 6 (p. 25)

5. **a** Quantity of one 'part' of paint = 12 ÷ 3 = 4 litres (✓)
   So 2 × 4 = 8 litres of red paint are needed. (✓)
   b Total quantity of orange paint made = 12 + 8 = 20 litres (✓)

Levels 5–7 (p. 28)

1. **a** 0.36 (✓)   **b** 0.1 (✓)   **c** 0.08 (✓)

2. **a** $\frac{1}{2} + \frac{1}{4} + \frac{1}{8} = \frac{4}{8} + \frac{2}{8} + \frac{1}{8} = \frac{7}{8}$
   So fourth piece is $1 − \frac{7}{8} = \frac{1}{8}$ of cake (✓)
   b $\frac{1}{4} = \frac{4}{16}$ (✓)
   So weight of $\frac{1}{4}$ of cake = weight of $\frac{1}{16} \times 4$
   = 20 × 4 = 80 g (✓)

3. **a** Percentage of pupils in Year 9
   $= \dfrac{(90 + 70)}{295} \times 100 = 54.2\%$ (✓)
   b Boys : girls = 150 : 145 (✓)
   = 150 ÷ 150 : 145 ÷ 150 = 1 : 0.967 (✓)

4. **a** $1 − \frac{3}{8} = \frac{8}{8} − \frac{3}{8} = \frac{5}{8}$ (✓)
   b $\frac{5}{8} \times 100 = 62\frac{1}{2}\%$ (✓)
   c $\frac{3}{8} \times 24 = £9$ (✓)

Level 8 (p. 29)

5. **a** Value of computer after 2 years
   $= £800 \times \dfrac{80}{100} \times \dfrac{80}{100}$ (✓) = £512 (✓)
   b Value of computer when new
   $= £704 \times \dfrac{100}{80} \times \dfrac{100}{80}$ (✓) = £1100 (✓)
   c **i** 0.8^2 or 0.64 (✓)   **ii** 0.8^n (✓)
   d Value after 4 years = £1500 × 0.8^4 (✓) = £614.40 (✓)
   e $0.8^3 = 0.512$ ($>\frac{1}{2}$), $0.8^4 = 0.4$ ($<\frac{1}{2}$)   (✓ *for method*)
   So it will take 4 years for the value of any computer to fall by half its original value.

168

PRACTICE QUESTION ANSWERS

At appropriate stages the marks to be awarded are shown as (✓).

3 Estimating

Levels 3–5 (p. 33)

1. 350 000 or 346 000 (✓)
2. **a** $x \div y = 5$ where $5 \times y = x$ (✓), e.g. where $x = 10, y = 2$
   b $24 \div 3 + 1 = 9$ (✓)
3. Any two of 2, 36; 3, 24; 4, 18; 6, 12; 8, 9 (2✓)
4. **a** e.g. 8, 9 (✓)   **b** e.g. 15, 3 (✓)
5. **a** $3 + 6 + (2 \times 4)$ (✓)   **b** $(6 + 4 + 2) \times 3$ (✓)
6. 7 (2✓)
7. 9 (2✓)
8. **a**

School	Number of pupils	To the nearest 100	To the nearest 10
Olney	884	900	880
Mesnes	662	700	660
Heaton	788	800	790
Pendle	906	900	910

   (4✓)
   b Any two numbers between 650 and 654 inclusive. (2✓)
9. **a** $4667 \div 359 \,(=13)$ or $4667 \div 13 \,(=359)$ (✓)
   b $1138 - 377 \,(=761)$ or $1138 - 761 \,(=377)$ (✓)
   c $1830 - 1103 \,(=727)$ or $727 + 1103 \,(=1830)$ (✓)
   d $17 \times 311 \,(=5287)$ or $5287 \div 311 \,(=17)$ (✓)
10. 37 (✓)

Levels 5–7 (p. 38)

1. **a** Cost of 4 garden pots $£2.45 \times 4 = £9.80$ (✓)
   b $12 \div 2.45 = 4.89$
      So you can buy 4 pots with £12.00 (✓)
   c $12 \div 4.49 = 2.67$
      So you can buy 2 pairs of pots with £12.00 (✓)
   d $£4.49 \times 3$ pairs $+ £2.45 \times 1$ single $= £15.92$
      So 7 pots is the greatest number you can buy with £16.00 (✓)
2.
Number	Rounded to 1 sf	Rounded to 2 sf	Rounded to 3 sf
0.5182	0.5	0.52	0.518
10.099	10	10	10.1
58.42	60	58	58.4
3486	3000	3500	3490

3. $\dfrac{81 \times 155}{42.4 \times 2.4} \approx \dfrac{80 \times 200}{40 \times 2}$ (1 sf) (✓) $= 200$ (✓)

Level 8 (p. 39)

4. **a** 5.115 (✓)   **b** 5.1249 or 5.125 (✓)
5. $\sqrt{\dfrac{12.2^3 \times 14.3}{440 \times 9.6^2}} \approx \sqrt{\dfrac{10^3 \times 10}{400 \times 10^2}}$ (1 sf) (✓) $= \dfrac{1}{2}$ (✓)

4 Patterns and sequences

Levels 3–5 (p. 43)

1. **a** Either 16×2 or 32×1 (✓)
   b $32 \div 6$ does not go exactly; 6 is not a factor of 32 (✓)
2. **a** (✓)
   b $22 \div 2 \times 5 = 55$ (✓)
   c Number of white tiles in odd pattern
      $= ((\text{Length} + 1) \div 2 \times 5) - 2$ or
      $((\text{Length} - 1) \div 2 \times 5) + 3$ (✓)
3. **a** 10, 15, 21 (✓)
   b Two prime numbers (e.g. 2, 3, 11, 13, 17, 19, …) (✓)
   c 15 is not prime, it has 2 factors; 3 and 5 as well as 1 and 15 (✓)

Level 6 (p. 44)

4. **a** 12 grey, 9 red (✓)   **c** $T = 4N - 3$ (✓)
   b 28 grey, 29 red (✓)

Levels 5–7 (p. 47)

1. **a** 49 grey, 4 blue (✓)
   b $14^2 = 196$ grey, 4 blue (✓)
   c n^2 grey, 4 blue (✓)
   d $n^2 + 4$ (✓)
   e $n^2 + n^2 + 3$ (✓) $= 2n^2 + 3$ (✓)

5 Formulas, expressions and equations

Levels 3–5 (p. 52)

1. **a** x^2 (✓)   **c** $x \times 2, x + x$ (✓)
   b $x \div 3$ (✓)   **d** e.g. $4x, 2x + 2x$ (✓)
2. **a** $2 \times n$ (✓)   **b** $2n - 5$ (✓)
3. **a** $n + n \ldots + n + n = 12n$ (✓)
   b $12n = 60$ (✓) so $n = 5$ (✓)

169

PRACTICE QUESTION ANSWERS

Level 6 (p. 53)

4 a Sapna : $3m - 4$ (✓)   Dale : $2m + 1$ (✓)
  b $3m - 4 = 2m + 1$ (✓)
  c $3m - 4 = 2m + 1$
     $3m - 2m = 1 + 4$
     $m = 5$ (✓)

5

x	y	
1	-1	too low
2	3	too high
1.5	0.75	too high (✓ for attempt between 1 and 2)
1.4	0.36	too high
1.3	-0.01	too low
1.35	0.1725	too high (✓ for attempt at 1.3 and 1.4)

The value which gives the solution nearest to 0 $x = 1.3$ (1 dp) (✓)

Levels 5–7 (p. 58)

1 $x + y$; $8a$; $-5q$; $5x + 5y - 4w$; $-4k$; $4x^2$ (3✓)

2 a 24 °C (thermometer shown)
  b $T < 24$ (✓)
  c $-9 \leqslant T < 12$ (✓)

3 a $P = \dfrac{x^2(x-1)}{4}$
     $P = \dfrac{5^2(5-1)}{4} = \dfrac{25 \times 4}{4} = 25$ (✓)
  b $Q = \dfrac{5y^3}{6}$
     $Q = \dfrac{5 \times 3^3}{6} = \dfrac{5 \times 27}{6} = 22.5$ (✓)

4 $6x - 2 = 4x + 5$
   $6x - 4x = 5 + 2$
   $2x = 7$
   $x = 3.5$ (✓)
   Area of square $= (4x + 5)^2 = (4 \times 3.5 + 5)^2$
     $= 19^2 = 361 \text{ cm}^2$ (✓)
   or
   Area of square $= (6x - 2)^2 = (6 \times 3.5 - 2)^2$
     $= 19^2 = 361 \text{ cm}^2$

Level 8 (p. 59)

5 a $y \geqslant 2$ (✓) $x \leqslant 2$ (✓) $y \leqslant x + 1$ (✓)
  b (graph showing lines $x=1$, $x=2$, $x=3$, $y=x+1$, $y=x-1$, $y=3$, $y=2$, $y=1$ with shaded region)

6 a $2(x - 3) - (x - 7) = 2x - 6 - x + 7$
              $= x + 1$ (✓)
  b $(x - 2)(x + 3) = x^2 + 3x - 2x - 6$
              $= x^2 + x - 6$ (✓)
  c $(x + 5)^2 = x^2 + 5x + 5x + 25$
          $= x^2 + 10x + 25$ (✓)

7 a $5 - 3x = 6x + 11$
     $-3x - 6x = 11 - 5$
     $-9x = 6$
     $x = -\dfrac{6}{9} = -\dfrac{2}{3}$
  b $2(x - 1) = 6$
     $2x - 2 = 6$ (✓)
     $2x = 8$
     $x = 4$ (✓)
  c $3x = 5(x - 1)$
     $3x = 5x - 5$
     $5 = 2x$
     $x = 2\tfrac{1}{2}$ (✓)

At appropriate stages the marks to be awarded are shown as (✓).

6 Functions and graphs

Levels 3–5 (p. 63)

1 a $A(2, 0)$, $B(3, 1)$, $C(4, 2)$
      (2✓ for all 3 correct, ✓ for 1 or 2 correct)
  b $(11, 9)$ (✓)
     The x co-ordinate is one more than the triangle number.
     The y co-ordinate is one less than the triangle number. (✓)
  c There needs to be a difference of two between the x and the y co-ordinates.
     $(x - y = 10 - 9 = 1 \neq 2)$ (✓)

2 a (graph showing PQRS with P(1,1), Q(1,2), R(3,2), S(3,1))   (2✓)
  b Square (✓)
  c $P'(0, 0)$, $Q'(0, 6)$, $R'(6, 6)$, $S'(6, 0)$
     (graph showing P'Q'R'S')   (2✓)
  d $36 \div 4 = 9$ times larger
     or (linear scale factor)$^2 = 3^2 = 9$ times larger (✓)

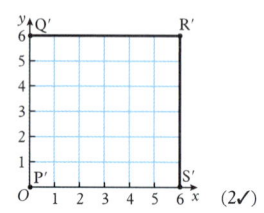

170

PRACTICE QUESTION ANSWERS

Level 6 (p. 64)

3 a, b

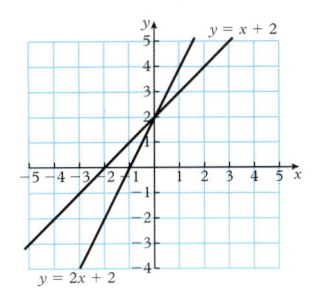

[Graph showing lines $y = x + 2$ and $y = 2x + 2$]

c $y = nx$ (e.g. $y = -2x, y = x, y = \frac{1}{3}x$, etc.) (✓)
d $y = x + 7$ (✓)

Levels 5–7 (p. 70)

1 a $y = 3$ (✓), $x = 1$ (✓)

b $y - x = 2$ is the equation of the line through D and B. (✓)

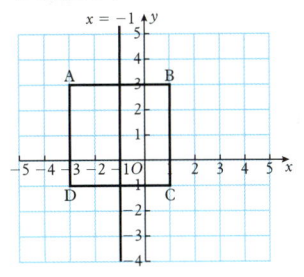

c [Graph with square ABCD and line $x = -1$]

One of $y = 1$ or $y - x = 2$ or $y = -x$ (✓)

d $2y + x = 22$ [1]
$y - x = 2$ [2]
$3y = 24$ [1] + [2]
$y = 8$ (✓)
Putting $y = 8$ in equation [2] gives
$8 - x = 2$
$x = 8 - 2 = 6$ (✓)

e (6, 8) i.e. answers from part **d**

2 $8x + 6y = 10$ [1]
$30x - 6y = 9$ [2] (✓ for rearrangement)
$38x = 19$ ADD [1] + [2]
$x = 0.5$ (✓)
Putting $x = 0.5$ in equation [1] gives
$4 + 6y = 10$
$6y = 6$
$y = 1$ (✓)

3 $x = 2$ (✓), $y = 1$ (✓)

Level 8 (p. 71)

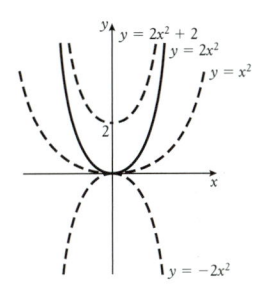

4 $y = 2x^2 + 2$ (✓); $y = x^2$ (✓); $y = -2x^2$ (✓)

[Graph showing parabolas $y = 2x^2 + 2$, $y = 2x^2$, $y = x^2$, and $y = -2x^2$]

5 i acceleration/increasing speed (✓)
ii level/same or constant speed (✓)
iii constant deceleration/braking (✓)

7 2D and 3D shapes

Levels 3–5 (p. 78)

1 a, b e.g.:

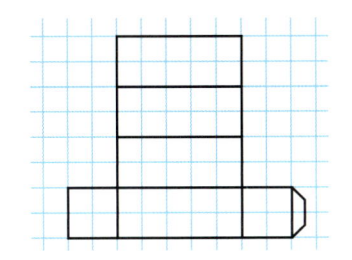

[Net of cuboid with flap]

(3✓ for net of cuboid, ✓ for flap)

2 a A (✓) **b** C (✓) **c** B, E (✓) **d** D (✓)
e Pupil's drawing of angle between 90° and 180° (✓)

3 a N (✓) **b** S (✓)

Level 6 (p. 79)

4 a $a = 180° - 70° = 110°$ (✓)
$b = 180° - 120° = 60°$ (✓)
$c = 180° - 70° - b° = 180° - 70° - 60° = 50°$ (✓)
b $d = 50°$ (alternate angles) (✓)
$e = 140°$ (alternate angles) (✓)
$f = 180° - 140° = 40°$ (angles on a straight line) (✓)
Trapezium (✓)

5 2.1 cm (3✓)

6 angles of 60° (✓), lengths correct to within 1 mm (✓)

171

PRACTICE QUESTION ANSWERS

Levels 5–7 (p. 84)

1. C (✓)
2. **a** $x = 360° \div 6 = 60°$ **b** $y = 60° + 90° = 150°$
3. Using Pythagoras
   Base $= \sqrt{(13^2 - 5^2)}$ (✓)
   $= \sqrt{(169 - 25)} = \sqrt{144} = 12$ cm (✓)
   Area $= \frac{1}{2}$ (base × height)
   $= \frac{1}{2} (12 \times 5) = 30$ cm^2 (✓)

Level 8 (p. 85)

4. **a** Using Pythagoras
   BC $= \sqrt{(450^2 + 350^2)}$ (✓) $= \sqrt{325\,000} = 570$ m (✓)
   b To find the bearing, b, from B to C first find angle x.

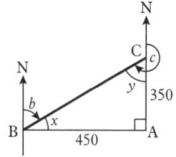

   Using trigonometry
   $\tan x = \dfrac{350}{450} = 0.777\ldots$
   $x = \tan^{-1} 0.777\ldots = 37.9°$ (3 sf)
   Therefore $b = 90° - 37.9° = 52.1°$
   Bearing from B to C is 052.1° (✓)

   c To find the bearing, c, from C to B first find angle y.
   $y = 180 - 90 - 37.9 = 52.1°$ (angles in a triangle)
   or $y = 52.1°$ (alternate angles)
   Therefore $c = 180° + 52.1° = 232.1°$
   Bearing from C to B $= 232.1°$ (✓)

   d $\cos 30 = \dfrac{450}{d}$ (✓)
   $d = \dfrac{450}{\cos 30} = 520$
   Distance from B to D $= 520$ m (✓)

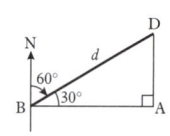

8 Position and movement

Levels 3–5 (p. 91)

1. **a** (✓) **b** (✓) **c** (✓) **d** (✓) **e** (✓) **f** Angle of rotation $= 90°$ (✓)

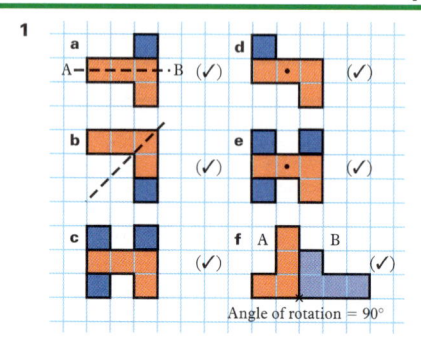

2. **a** (✓) **b** (✓) **c** (✓)
3. (2✓)

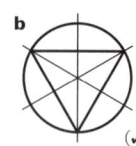

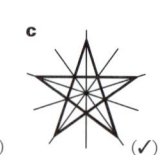

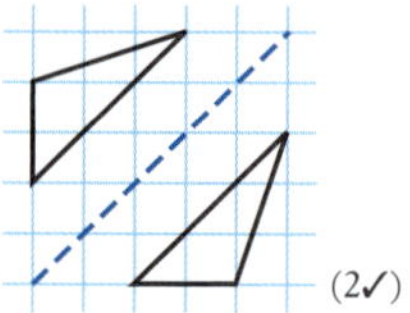

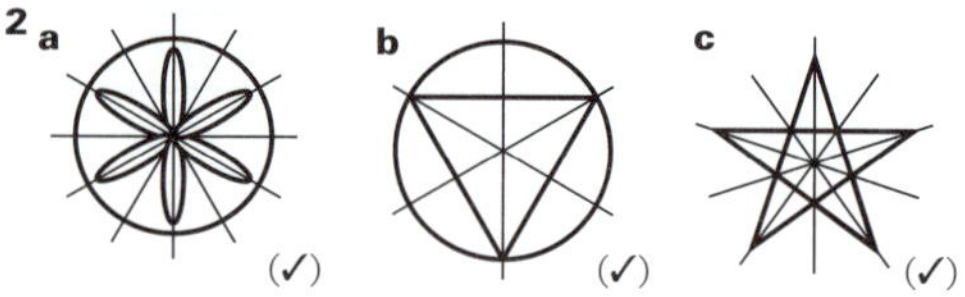

Level 6 (p. 92)

4. **a** 12 (✓) **b** 2 (✓)
5. Circle drawn (✓)
   Trapezium drawn (✓)
   Square drawn (✓) 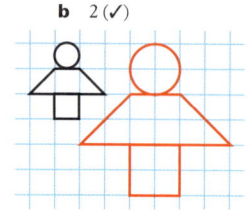
6. Correct size (×3) (✓)   Correct position (✓)

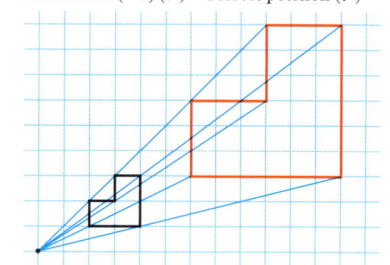

Levels 5–7 (p. 97)

1. **a** $42 \div 10 = 4.20$
      $30 \div 7 = 4.29$
      Scale factor $= 4.20$ (2 dp) (✓)
   b $30 \div 10 = 3.00$
      $21 \div 7 = 3.00$
      Scale factor $= 3.00$ (2 dp) (✓)
   c $21 \div 10 = 2.10$
      $15 \div 7 = 2.14$
      Scale factor $= 2.10$ (2 dp) (✓)

2. (✓ for drawing rectangle 6 cm × 3 cm)
   a Line 1.5 cm from side BC (✓)
   b Line at 45° to side DC (✓)
   c Shaded $\frac{1}{4}$ circle radius 1 cm, centre A (✓)

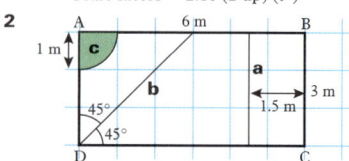

172

PRACTICE QUESTION ANSWERS

3   Scale 1 cm : 1 m

(4 ✓)

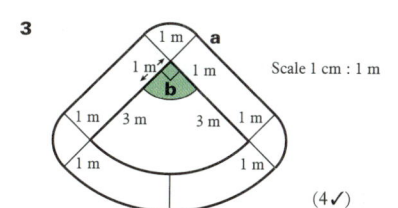

Level 8 (p. 98)

4   $\dfrac{h}{70} = \dfrac{84}{60}$ (✓)

So $h = \dfrac{84 \times 70}{60} = 98$ mm (✓)

or scale factor = 70 ÷ 60 (✓)
so $h = 84 \times 70 ÷ 60$ (✓)

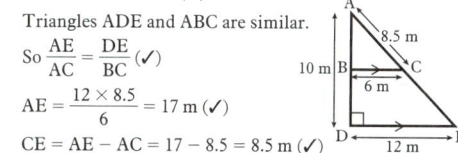

5   Triangles ADE and ABC are similar.
So $\dfrac{AE}{AC} = \dfrac{DE}{BC}$ (✓)

$AE = \dfrac{12 \times 8.5}{6} = 17$ m (✓)

CE = AE − AC = 17 − 8.5 = 8.5 m (✓)

6   a   ASA (✓)
    b   BAC = 90° (✓); AC = DE = $\sqrt{(8^2 - 6^2)}$ = 5.29 cm (2 ✓)

7   a   A, D (✓); enlargement only (no sides the same) (✓)
    b   A, E (✓); AAS (✓)
    c   C, D (✓); AAS (✓)

At appropriate stages the marks to be awarded are shown as (✓).

9 Units of measurement

Levels 3–5 (p. 104)

1   a   10 : 20 (✓)   b   09 : 45 (✓)   c   35 minutes (✓)
    d   07 : 30 or 08 : 15 or 09 : 00 (✓)
    e   08 : 15 (✓)
    f   (✓)

2   a   6.4 cm (✓) 64 mm (✓)
    b   16 mm = 1.6 cm
        Length of pencil = 5.3 − 1.6 = 3.7 cm (✓)

3   a   2.2 kg (✓)
    b   600 g = 0.6 kg (✓)
        Mass of cheese = 2.2 − 0.6 = 1.6 kg (✓)

4   a   1 kg (✓)
    b   Value between 0.5 and 0.7 litres (✓)
    c   Value between 4 and 9 mm (✓)

5   a   Value between 4.9 and 5.1 cm (✓)
    b   Value between 7.7 and 7.9 in (✓)

Levels 5–7 (p. 108)

1   a   7.2 cm (✓)
    b   Value between 2.6 and 2.8 kg (✓)
    c   Value between 5.32 m and 5.34 m (✓)

2   a   12 st 10 lb (✓)
    b   14 st 3 lb (✓)

3   a
Time (min)	1	6	10	15	25	30	50
Capacity (litres)	12	72	120	180	300	360	600

(2 ✓)

    b $C = 12T$ (✓)
    c $T = \dfrac{C}{12} = \dfrac{5000}{12} = 416.666\ldots$ minutes (✓)
        416.666 … minutes = 6 hours 57 minutes
        (rounding to nearest minute)

4   a   £550 = 880 euro
        So £1.00 = $\dfrac{880}{550}$ = 1.60 euro (✓)
    b   26.40 ÷ 1.60 = £16.50 (2 ✓)

5   a   10 minutes (✓)
    b   Adam and Belinda passed each other going in opposite directions (✓)
    c   50 minutes = $\dfrac{50}{60} = \dfrac{5}{6}$ hours

        Average speed = $\dfrac{\text{Distance}}{\text{Time}}$

        Average speed = $\dfrac{10}{\frac{5}{6}} = \dfrac{10 \times 6}{5}$ = 12 km/h (✓)

    d   Time Belinda was moving = 20 + 15 + 10
        = 45 minutes = $\frac{3}{4}$ hours

        Average speed = $\dfrac{10}{\frac{3}{4}} = \dfrac{10 \times 4}{3}$ = 13.3 km/h (✓)

    e   10 and 20 minutes (✓)

10 Perimeter, area and volume

Levels 3–5 (p. 113)

1   a   i   Perimeter = 10 cm (✓)   ii   Area = 6 cm^2 (✓)
    b   i   Perimeter = 18 cm (✓)   ii   Area = 8 cm^2 (✓)

2   a   7 (✓)   b   10 (✓)

3   a   Any shape (other than the one shown) with 6 tiles connected correctly. (✓)

4   a $5 \times 3 = 15$ cm^2 (✓)
    b $\frac{1}{2} \times 6 \times 8 = 24$ cm^2 (✓)
    c $\frac{1}{2} \times 4 \times 7 = 14$ cm^2 (✓)

5   a $12 \times 2 = 24$ (✓)   b $6 + (12 \times 2) = 30$ (✓)

Level 6 (p. 114)

6   a   Circumference = $\pi \times$ diameter
        = $\pi \times 12 = 37.69$ mm (✓)
        Distance moved = 2 × circumference
        = 2 × 37.69 = 75.4 mm (3 sf) (✓)

173

PRACTICE QUESTION ANSWERS

   b Number of turns = $\dfrac{\text{Distance}}{\text{Circumference}}$

                    $= \dfrac{201}{\pi \times 16} = \dfrac{201}{50.27}$

                    $= 3.998 = 3.40$ (3 sf) (✓)

7 Length of side = $\sqrt{9} = 3$
     Perimeter = $3 \times 4 = 12$ cm (✓)
8 $7 \times 4 = 28$ cm² (✓)
9 $\tfrac{h}{2}(7 + 5) = 24$
     $12h = 48$, so $h = 4$ (✓)

Levels 5–7 (p. 118)

1 **a** Area of shaded face = $\tfrac{1}{2} \times 7 \times 8$ (✓) = 28 m² (✓)
    b Volume of prism = Area of base × Height
         = $28 \times 12 = 336$ m³ (✓)

2 **a** Area of semicircle = $\tfrac{1}{2}\pi r^2 = \tfrac{1}{2}(\pi \times 5^2)$
         = 39.27 cm² (2 dp) (✓)
    b Perimeter of semicircle = $\tfrac{1}{2}\pi d + d = 15.708 + 10$
         = 25.71 cm (2 dp) (✓)

3 $\dfrac{h}{60} = \dfrac{20}{100}$ (✓) so $h = \dfrac{20 \times 60}{100} = 12$ cm (✓)
    or cross-section area = $100 \div 20 = 5$ cm² (✓)
    $5h = 60$ so $h = 12$ cm (✓)

4 If width is x cm, then length is $3x$ cm and perimeter is
    $3x + x + 3x + x = 8x$ cm
    $8x = 48$ cm so $x = 6$ cm
    Area of rectangle = $x \times 3x = 3x^2$
         = $3 \times (6^2) = 108$ cm² (✓)

5 $36 \div 2 = 18$, $100 \div 2 = 50$, $50 \div 2 = 25$ (✓)
    Number of cubes that will fit in box = $18 \times 50 \times 25$
         = 22 500 (✓)

6 Area of shaded region = $\pi(3^2) - \pi(1.2^2)$ (✓)
         = $28.27 - 4.52 = 23.75$ (2 dp) (✓)

7 Surface area of prism = $(\tfrac{1}{2} \times 5 \times 12) + (\tfrac{1}{2} \times 5 \times 12)$
     + $(13 \times 10) + (12 \times 10) + (5 \times 10) = 360$ cm² (✓)

Level 8 (p. 119)

8 Surface area of curved face = $\tfrac{1}{2}(4\pi r^2) = 2\pi r^2$ (✓)
    Surface area of face = πr^2 (✓)
    Surface area of hemisphere = $2\pi r^2 + \pi r^2 = 3\pi r^2$
         = $3\pi 4^2 = 48\pi$
         = 150.8 cm² (4 sf) (✓)

9

$4\pi r^2$	lbh	$\tfrac{1}{2}bh$	$\tfrac{4}{3}\pi r^3$	$4(r+b)$	$\dfrac{\pi h^2}{b}$	
				✓	✓	Length
✓		✓				Area
	✓		✓			Volume

                                      (2✓)

At appropriate stages the marks to be awarded are shown as (✓).

11 Units of measurement

Levels 3–5 (p. 127)

1 **a** Monday 28 (✓) Tuesday 22 (✓) Wednesday 15 (✓)
    b Thursday: $2\tfrac{1}{4}$ discs (✓) Friday: 6 discs (✓)

    c Thursday (✓)
    d The pictogram summarises the data and shows it in such a way that it is easily understood. Makes it easy to compare the figures. (✓)
    e A tally-chart is easy to add to, and to use to collect the data. (✓)

2 **a** Number of programmes recorded
         = $16 + 11 + 12 + 13 = 52$ (✓)
    b Number of films recorded = $16 + 12 = 28$ (✓)
    c

	Channel 1	Channel 3
Soaps	7	10
Comedy	4	6

                                     (2✓)

3 **a** Dog (✓) **b** Fish and mice (✓)
    c Number of cats = $40 \times \tfrac{1}{4} = 10$ (✓)
    d Number of dogs = number between 12 and 15 (✓)

Level 6 (p. 129)

4 **a** There are too many tallies. The tallies needed to be counted up to make the chart clearer. (✓)
    b 180 (✓)
    c A pie-chart will show better a comparison between the holiday choices, as a fraction of those questioned. It is not how many who have made a choice that is important, but the proportion of each that have shown a preference. (✓)
    d Ski 26°, Winter abroad 70°, Summer sun 174°, Touring 44°, UK holiday 46°
         (2 angles correct ✓; all angles correct 2✓; labels and title ✓)

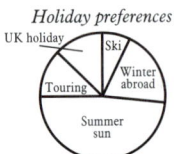

Holiday preferences

Levels 5–7 (p. 132)

1 **a** Between 30% and 40% (✓)
    b Between 15 and 25 (✓)
    c There are many more items included in B than A (✓)
    d (2 angles correct ✓; all angles correct 2✓; labels and title ✓)

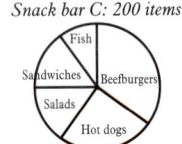

Snack bar C: 200 items

2 **a** Any three from:
       Amounts overlap
       No 'don't know' box
       Amounts are too small
       Groupings need to be bigger to cover more amounts (3✓)
    b Need to include both girls and boys, and pupils of different ages/year, groups and backgrounds. (2✓)

3 **a** There is little change in Gary's figures. Therefore his week 11 sales are likely to be similar to his sales in weeks 1–10.
    b Though Ali's sales figures in some weeks are higher than Gary's, in many more weeks they are lower than Gary's. Therefore, Ali's overall sales are lower.

PRACTICE QUESTION ANSWERS

12 Averages and spread

Levels 3–5 (p. 138)

1. a Mode = 2 (✓)
    b Range = 6 − 1 = 5 (✓)
    c Median = 3 (✓)
    d Mean = 33 ÷ 10 = 3.3 (✓)

2. To score an average of 12, Jerry must score $3 \times 12 = 36$ over three games. In 2 games he has scored $8 + 15 = 23$ points. So he must score $36 - 23 = 13$ points in his final game.

3. a Mean = $\dfrac{\text{Total number of pets}}{\text{Total number of pupils}}$

    $= \dfrac{(5 \times 0) + (3 \times 1) + (2 \times 2) + (2 \times 3) + (3 \times 4) + (2 \times 5)}{5 + 3 + 2 + 2 + 3 + 2}$ (✓)

    $= \dfrac{35}{17} = 2.06$ (3 sf) (✓)

    b Modal number = 0 pets (✓)
    c Median number is 9th value = 2 pets (✓)

Level 6 (p. 139)

4. a Mean number of errors = $\dfrac{\text{Total number of errors}}{\text{Total number of pupils}}$

    $= \dfrac{(1 \times 5) + (2 \times 6) + (3 \times 7) + (4 \times 5) + (5 \times 3)}{5 + 6 + 7 + 5 + 3}$ (✓)

    $= \dfrac{73}{26} = 2.81$ (3 sf) (✓)

    b Median = 13th value = 3 errors (✓)
    c The mean, as it is the result of a calculation, need not be one of the given number of errors. The median is a given number of errors. (✓)

5. a As the temperature rises, the number of cups of hot soup decreases (inverse proportion, negative correlation). (✓)
    b No relation (no correlation). (✓)

Levels 5–7 (p. 145)

1. a As the number of hours of clear sky increases, the temperature decreases (inverse proportion, negative correlation). (✓)
    b Line of best fit drawn ±2 mm (✓)
    2 °C → 4.8 h ± 1 h (✓)
    4 h → 3 °C ± 1 °C (✓)

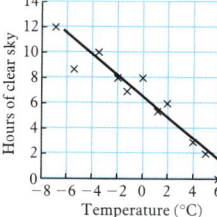

2. a Mean = $\dfrac{\sum fx}{\sum f}$

    $\sum fx = (34.5 \times 3) + (44.5 \times 8) + (54.5 \times 11) + (64.5 \times 9) + (74.5 \times 13) + (84.5 \times 6) = 3115$ (✓)

    Mean = $\dfrac{3115}{50} = 62.3$ paces (✓)

    b Median is 25th → group 60–69 (✓)

Level 8 (p. 147)

3. a Median age = 47 ± 2 (✓)
    b Interquartile range: 150 on vertical axis → 57 ± 2
    50 on vertical axis → 36 ± 2
    Interquartile range = 57 − 36 = 21 (2 ✓)
    c Percentage of population under 50 ≈ $\dfrac{116}{200} \times 100$
    = 58% ± 2% (✓)

13 Probability

Levels 3–5 (p. 151)

1. a $\frac{2}{5}$ (✓)   b 0 (✓)
    c Probability of getting a blue or brown cube is $\frac{2}{5} = 0.4$ (✓)

    d Probability of **not** getting a green cube is $1 - \frac{1}{5} = \frac{4}{5} = 0.8$ (✓)

2. a Probability of red = $\frac{3}{8}$
    Probability of green = $\frac{3}{8}$
    So red is no more likely than green. (✓)
    b The wheel is just as likely to stop on yellow the second time as it was the first.
    So probability of yellow = $\frac{1}{8}$ (✓)
    c All segments one colour. (✓)
    d Any four segments one colour. Remaining four segments a second colour. (✓)

3. a Coin is just as likely to land on heads the fifth time as it was the other four times. So probability of heads = $\frac{1}{2} = 0.5$

    b Probability of 2 heads = $\frac{1}{2} \times \frac{1}{2} = \frac{1}{4}$ (✓)
    c Probability of not throwing 2 tails
    = 1 − probability of 2 tails = $1 - \frac{1}{4} = \frac{3}{4}$ (✓)

PRACTICE QUESTION ANSWERS

Level 6 (p. 152)

4 a It is possible that the bag contains many different coloured counters and that it is only by chance that she has picked only red counters. (✓)

b The probabilities of picking counters from the bag depend on the colours of the counters in the bag, not on the colours of the counters already taken out of the bag. (✓)

Levels 5–7 (p. 56)

1 a

+	2	4	6	8
1	3	5	7	9
3	5	7	9	11
5	7	9	11	13
7	9	11	13	15

(✓)

b Probability that answer is greater than $9 = \frac{6}{16} = \frac{3}{8}$ (✓)

c Probability that answer is a square number = Probability that answer is $9 = \frac{4}{16} = \frac{1}{4}$ (✓)

d Probability that answer is an even number = 0 (✓)

2 a Estimated probability of red $= \frac{30}{60} = \frac{1}{2}$ (✓)

b 10 times (✓)

c Probability only tells you what is likely to happen, not what actually happens. (✓)

d To improve the accuracy of his results, Tony needs to do more trials, i.e. throw the cube more times. (✓)

Level 8 (p. 57)

3 a Events are independent. So probability of passing through both sets of lights without having to stop $= 0.3 \times 0.3 = 0.09$ (✓)

b Probability of being stopped at only one set of lights $= (0.7 \times 0.3) + (0.3 \times 0.7)$ (✓)
$= 0.21 + 0.21 = 0.42$ (✓)

c Probability of being stopped at first set of lights = 0.7 Number of times Karen can expect to be stopped in 80 journeys $= 0.7 \times 80 = 56$ times

Test Tips

In the days before the tests

- Read through the 'What you need to know?' sections for your tier again.

- Read the Test Yourself tests and the Practice Questions for your tier of entry. Check the answers too and make sure that you know how to do all of the questions you have seen.

- Ask your teacher for copies of the Maths Tests from last year and the year before. If you try questions from these papers, ask your teacher for the answers and check your work.

On the day of the tests

Take the right equipment with you: pen, pencil, rubber, ruler, protractor or angle measurer, compasses and a calculator for the calculator paper and anything else that your teacher has told you to bring.

In the examination room

- Read each question carefully. Make sure you understand what the question is about and what you are expected to do.

- In all questions show any working out you do, even if you are using a calculator. You may lose marks if you do not show working.

- When using rulers, protractors or compasses make sure you use them accurately.

- The number of marks for each question is always shown on the paper.

- Give your answers to an appropriate degree of accuracy. **Do not** make them inaccurate by rounding off.

- Always round money answers to the nearest penny and to a given number of significant figures **where this is asked for**.

- If you find a question too hard, go on to the next question. But try to write something for each part of every question.

- If you have spare time at the end, use it wisely to check over your answers and working.